Biophysical Thermodynamics
of Intracellular Processes

Lev A. Blumenfeld   Alexander N. Tikhonov

Biophysical Thermodynamics of Intracellular Processes

Molecular Machines of the Living Cell

With 54 Illustrations

Springer-Verlag

New York Berlin Heidelberg London Paris
Tokyo Hong Kong Barcelona Budapest

Lev A. Blumenfeld
Institute of Chemical Physics
Russian Academy of Sciences
Kosygin Str. 4, V-334
Moscow 117977
Russia

Alexander N. Tikhonov
Department of Biophysics
Faculty of Physics
M.V. Lomonosov State University
Moscow 119899
Russia

Library of Congress Cataloging-in-Publication Data
Blumenfeld, L.A. (Lev Aleksandrovich)
    Biophysical thermodynamics of intracellular processes: molecular
machines of the living cell / Lev A. Blumenfeld, Alexander N.
Tikhonov.
        p. cm.
    Includes bibliographical references and index.
    ISBN 0-387-94179-7
    1. Thermodynamics. 2. Bioenergetics. 3. Enzyme kinetics.
I. Tikhonov, A.N. (Aleksandr Nikolaevich) II. Title.
QP517.T48B58 1994
574.19'121—dc20   93-35826
                                                              CIP

Printed on acid-free paper.

Production coordinated by Brian Howe and managed by Terry Kornak; manufacturing super-
vised by Gail Simon.
Typeset by Asco Trade Typesetting Ltd., Hong Kong.
Printed and bound by Braun-Brumfield, Inc., Ann Arbor, MI.
Printed in the United States of America.

9 8 7 6 5 4 3 2 1

ISBN 0-387-94179-7 Springer-Verlag New York Berlin Heidelberg
ISBN 3-540-94179-7 Springer-Verlag Berlin Heidelberg New York

Preface

This book is aimed at a large audience: from students, who have a high-school background in physics, mathematics, chemistry, and biology, to scientists working in the fields of biophysics and biochemistry. The main aim of this book is to attempt to describe, in terms of physical chemistry and chemical physics, the peculiar features of "machines" having molecular dimensions which play a crucial role in the most important biological processes, viz., energy transduction and enzyme catalysis. One of the purposes of this book is to analyze the physical background of the high efficiency of molecular machines functioning in the living cell.

This book begins with a brief review of the subject (Chapter 1). Macromolecular energy-transducing complexes operate with thermal, chemical, and mechanical energy, therefore the appropriate framework to discuss the functioning of biopolymers comes from thermodynamics and chemical kinetics. That is why we start our analysis with a consideration of the conventional approaches of thermodynamics and classical chemical kinetics, and their application to the description of bioenergetic processes (Chapter 2). Critical analysis of these approaches has led us to the conclusion that the conventional approaches of physical chemistry to the description of the functioning of individual macromolecular devices, in many cases, appear to be incomplete. This prompted us to consider the general principles of living machinery from another point of view. In Chapter 3 we discuss a "machine-line" approach to the functioning of biopolymers, and consider some new results concerning, in particular, the models which illustrate the high efficiency of energy transduction by molecular machines, as well as the unusual thermodynamic behavior of small systems. A number of experimental and theoretical results, relevant to the problem of enzyme catalysis and energy transduction in biomembranes, are considered in Chapters 4 and 5. In Chapter 4 we analyze the earlier theories of enzyme catalysis, and describe the relaxation concept of enzyme catalysis which has been developed and considered at length by one of us in two previous books published by Springer-Verlag (L.A. Blumenfeld, *Problems of Biological Physics*, 1981, and *Physics of Bioenergetic Processes*, 1983). In this book, using new experimental results that have appeared since the above-mentioned books were published, we have tried to extend the experimental and theoretical principles for the relaxation concept of functioning "molecular machines." In Chapter 5 we consider the processes of energy transduction in biological membranes, focusing on ATP synthesis in chloroplasts. Since we wanted to draw attention mainly to the conceptual aspects of the problem, it was beyond our scope to analyze all the variety of new experimental results concerning different aspects of enzyme catalysis and bioenergetics. However, to provide experimental evidence for certain crucial points of our approach to enzyme catalysis and membrane phosphorylation, in Chapters 4 and 5 we appealed to experimental data obtained by different authors, as well as to recent experimental results obtained in their laboratories.

Contents

Preface v

CHAPTER 1
Introduction 1

CHAPTER 2
Thermodynamics and Chemical Kinetics of Living Systems 4
2.1. How Scientists Learned to Distinguish Energy from Force
     (Brief Historic Review) 4
2.2. Kinetics and Thermodynamics of Chemical Reactions 6
2.3. Applicability of Equilibrium and Nonequilibrium Thermodynamics to
     Biological Systems and Processes 16
2.4. The Mechanisms of Energy Coupling in Chemical Reactions 20
     2.4.1. Indirect Mechanism of Energy Coupling in Equilibrium
            (Quasi-Equilibrium) Homogeneous Mixtures of Chemical
            Reagents 21
            2.4.1.1. Enthalpic Mechanism of Indirect Coupling 22
            2.4.1.2. Entropic Mechanism of Indirect Coupling 24
     2.4.2. Entropic Mechanism of Coupling Chemical Reactions
            in Open Systems 30

CHAPTER 3
Molecular Machines: Mechanics and/or Statistics? 38
3.1. The Second Law of Thermodynamics and Its Application
     to Biochemical Systems 38
3.2. Energy-Transducing Molecular Machines 45
     3.2.1. Macroscopic Machines 45
     3.2.2. What Are Molecular Machines? Reversibility of
            Energy-Transducing Devices and the Problem of the Optimal
            Functioning of Molecular Machines 48
     3.2.3. Models for Calculating the Conversion Factor 52
3.3. Statistical Thermodynamics of Small Systems, Fluctuations, and the
     Violation of the Mass Action Law 60

3.3.1. Structural Peculiarities of Energy-Transducing Organelles of
Chloroplasts 61
3.3.2. Chemical Equilibrium Inside Small Vesicles 64
3.3.3. Compartmentalization and the Problem of the Macroscopic
Description of "Channeled" Chemical Reactions 73
3.3.4. The Fluctuations, Random Noise, Energy Transduction, and
Apparent Violation of the Second Law of Thermodynamics 77

CHAPTER 4
Principles of Enzyme Catalysis 86
4.1. Introduction 86
4.2. Earlier Theories of Enzyme Catalysis 89
4.3. The Relaxation Concept of Enzyme Catalysis 94
4.4. Protein Dynamics and Enzyme Functioning 101
4.4.1. Theoretical Aspects of Protein Structural Dynamics 102
4.4.2. Experimental Evidence for Protein Nonequilibrium States and
Their Evolution in the Course of Enzyme Turnover 106

CHAPTER 5
Energy Transduction in Biological Membranes 112
5.1. Introduction: Two Views on the Problem of Energy Coupling in
Biomembranes 112
5.2. Transmembrane Electrochemical Proton Gradients in Chloroplasts 121
5.2.1. Brief Review of the Methods for the ΔpH Measurements with
pH-Indicating Probes 121
5.2.2. Measurements of ΔpH in the Thylakoids with the Kinetic Method   123
5.2.3. Measurements of ΔpH in the Thylakoids with a Spin Labeling
Technique 127
5.2.4. Lateral Heterogeneity of ΔpH in Chloroplasts 133
5.2.5. Membrane-Sequestered Proton Pools and Alternative Pathways
of Proton Transport Coupled with ATP Synthesis 140
5.3. Mechanism of ATP Formation Catalyzed by H^+ATPsynthases 144
5.3.1. An Elementary Act of ATP Synthesis 144
5.3.1.1. Initial Events of ATP Formation 144
5.3.1.2. Energy-Requiring Step of ATP Formation 146
5.3.1.3. ATP Synthesis from ADP and P_i Catalyzed by
Water-Soluble Coupling Factor F_1 151
5.3.1.4. ATP Synthesis Induced by the Acid-Base Transitions 152
5.3.1.5. ATP Synthesis from ADP and P_i as Considered from the
Viewpoint of the Relaxation Concept of Enzyme Catalysis   154
5.3.2. ATP Synthesis under Steady State Conditions 159
5.3.2.1. The Possible Model for ATPsynthase Cyclic Functioning   159
5.3.2.2. Photophosphorylation in Chloroplasts and Oxidative
Phosphorylation in Mitochondria 162

Afterword 174

Index 177

CHAPTER 1

Introduction

What is life? In what aspects do living creatures differ from nonliving ones? The history of science cites numerous futile attempts to obtain answers to these questions. The practical impossibility of formulating an unambiguous scientific criteria of life concerned many great scientists. Erwin Schrödinger, in his famous book *What Is Life? The Physical Aspects of the Living Cell* [1], after considering the physical foundations of genetics, reached the following conclusion in the last chapter: the only real criterion of living matter, but unamenable to the scientific analysis, is the existence of individual consciousness (probably, nonmaterial and nondestructible). From the viewpoint expressed by one of us (L.A.B.) in [2], the problem of individual consciousness lies outside contemporary and even future science. However, in order to understand the structure and functioning of all biological objects it is quite enough to use the well-known principal laws of physics. This statement is a rather unprovable *"symbol of faith."*

It is necessary to underline that we are speaking about the *principal* laws of physics. The situation here is the same as for the description of a sophisticated electronic device. The behavior of such a construction can be specific enough. But it would be futile, for example, to search for new physical laws to describe the properties of any new scheme of a television set. These properties obey definite and complicated rules, but can be understood completely using the physical principles of the functioning of device elements and knowledge of the construction scheme.

There are, of course, other situations in science. We cannot understand magnetic and electric phenomena on the basis of mechanics without introducing new postulates, which are the generalization of experiments and could not be derived logically from the laws of mechanics. The introduction of new postulates is always connected to the appearance of universal constants (for electrical and magnetic phenomena, i.e., the velocity of light propagation, c). An adequate description of the laws of the microworld requires the introduction of quantum mechanical postulates and the universal constant "h" (Planck's constant).

1

The above-mentioned symbol of faith is, thus, the conviction that understanding the structure and functioning of living objects does not need the use of new fundamental postulates and universal constants. This does not mean, however, that in studying the living we do not face new physical problems. The advancement of biology is stimulated by the development of physical sciences. Objects and processes which confront the physicist or chemist studying living matter make it necessary for him to reconsider the limits of applicability of many conventional approaches. First of all, such reconsideration is determined by the fact that the main participants of important biological processes are macromolecules (biopolymers), every one of which, being a statistical system, reveals the properties of a construction.

Biological macromolecules are not just the constructions but the machines as well, i.e., constructions "designed" for the directional transfer and transduction of energy (see [3] regarding the teleological approaches to biology). In conventional macroscopic machines *statistical* and *mechanical* components are separated spatially. From the viewpoint of the physicist, quite different approaches have to be used for the analysis of mechanical and statistical devices. Let us consider, for example, a combustion engine. The statistical (thermodynamical) component of the engine is the fuel mixture inside the cylinders. Its behavior is described by the methods of chemical thermodynamics and chemical kinetics. Averaging out over the immense number of particles, one introduces the macroscopic parameters of a system such as temperature, pressure, entropy, etc. The mechanical components of the engine, i.e., pistons, hinges, and consoles, also have immense numbers of molecules, and thus possess a multitude of degrees of freedom (each individual molecule has several translational, vibrational, and rotational degrees of freedom). However, for each of these macroscopic parts there exist only few mechanical degrees of freedom completely determined by the system construction. The existence of at least one mechanical degree of freedom (e.g., the position of a movable piston in a cylinder of a heat engine) is the obligatory condition for mechanical machine functioning, i.e., the performance of work by means of the spatial transfer of energy.

For biopolymers, separation into statistical and mechanical components is impossible; the statistical and mechanical subsystems spatially coincide. An adequate description of such objects requires us to extend the conventional approaches of statistical thermodynamics that, in turn, could lead to the development of new parts of this science.

Molecular machines are very small. In many cases this property, as well as the relatively low values of energy liberated or absorbed in each individual cycle, lead to unexpected consequences. For example, there can be certain deviations from the well-established rules of the operation of macroscopic machines. Due to the small dimensions of the systems, one has to take into account the fluctuations of thermodynamic parameters. Relatively low-energy changes in each individual step lead to the necessity to account for additional energy losses to obtain information (see, for references, [4]).

The main aim of this book is the description of the peculiar features of machines having molecular dimensions, and which play a principal role in the most important biological processes: enzyme catalysis and energy transduction. All machines in biological systems are chemical machines. They were "designed" in the course of biological evolution in order to utilize energy liberated or absorbed by the intracellular chemical reactions. At the close of the nineteenth century and in the first half of the twentieth century there originated two branches of science dealing with the laws of chemical reactions: chemical thermodynamics and chemical kinetics. It is appropriate, therefore, to begin with an account of the main foundations of these sciences.

References

1. E. Schrödinger (1945), *What is Life? The Physical Aspects of the Living Cell*, Cambridge University Press, Cambridge.
2. L.A. Blumenfeld (1989), *Life and Science (USSR)*, No. 10, p. 60.
3. L.A. Blumenfeld (1981), *Problems of Biological Physics*, Springer-Verlag, Heidelberg.
4. L.A. Blumenfeld (1983), *Physics of Bioenergetic Processes*, Springer-Verlag, Heidelberg.

CHAPTER 2

Thermodynamics and Chemical Kinetics of Living Systems

2.1. How Scientists Learned to Distinguish Energy from Force (Brief Historic Review)

In the preceding pages of this book we used the word "energy." This word is one of the many routine terms in ordinary language to which people get accustomed from childhood, without dwelling on their exact meaning. The custom replaces understanding—a rather wide-spread phenomenon. Meanwhile, energy and the laws of energy transformation represent the very foundation of the whole science

The first man to use this term ("$\varepsilon\nu\varepsilon\rho\gamma\iota\alpha$" in Greek), without an exact scientific meaning, was probably Aristotle, more than 300 years B.C. The notion of energy under different terms appeared with the development of modern physics, in the fifteenth and sixteenth centuries, when the observations and their quantitative treatment (e.g., J. Kepler) were supplemented with deliberately set experiments (e.g., Galileo). Modern physics began with mechanics. The fundamental property of matter is motion. The generally accepted main philosophical principle from the past was the following statement: every action has a cause. The cause of motion was called *force*. Force could quite easily be associated with physical effort: everybody understands that physical effort can initiate motion. There is one other obvious notion connected with daily experience, i.e., *work*. To lift the same weight to different heights requires different work to be performed. The cause of work was also called force. The identical term applied to two essentially different concepts caused misunderstanding and sharp discussions, between the greatest scientists, over several decades.

These discussions were connected with the question: What is the measure of force? In the seventeenth century two scientists proposed two different answers. Let force "F" be applied to the body having mass "m" during time interval "t." At the end of this interval the velocity of a body will be "V." It was shown experimentally that the force is proportional to the mass, as well as to the acquired velocity. Thus, the force $2F$ applied to the mass m during time t leads to the velocity $2V$. The measure of force associated with the mechanical motion is thus mV. This quantity was later called momentum or impulse. The above reasoning is attributed to the great French scientist René Descartes.

The approach of the famous German scientist Leibnitz was also based on experimental data. These experiments were performed by Galileo at the close of the sixteenth century. He demonstrated that a body thrown upward reached a height which was proportional to the square of the initial velocity. It was quite clear, Leibnitz thought, that to lift weight m to height $4h$, one must spend the same "force" as to lift weight $4m$ to height h. But, according to Galileo's experiments, to lift a weight to height $4h$, the initial velocity must only be two times greater than to lift weight $4m$ to height h. The measure of force proposed by Descartes is therefore wrong

$$m * 2V \neq 4m * V.$$

The true measure of "*force*" (or "*living force*," using the Leibnitz terminology—"*Vis vita*" in Latin) is consequently mV^2. Indeed, $m * (2V)^2 = 4m * V^2$. The contemporary expression $mV^2/2$ was introduced by the French scientist Kariolis.

The creator of mechanics, the great scientist Isaac Newton, was totally on the side of Descartes. He suggested the well-known definition of force

$$F = mV/t = ma,$$

where a is acceleration. Newton continuously ignored the Leibnitz approach. Leibnitz, in turn, called the force according to Newton, "*Vis mortua*," a dead force. This terminological confusion significantly slowed down the formulation of the principle of energy conservation.

Due to the immense authority of Newton, the Leibnitz idea and the very notion of work performed by force were practically forgotten until the middle of the eighteenth century, when the Swiss scientist Bernoulli began to use the term "Vis vita" again. He understood that in many cases "Vis vita" apparently disappears, but the ability to perform work is preserved and only transfers into a different form, e.g., into the compression of a spring. Euler took one more step forward. He stated that the "Vis vita" of a material body is a one-valued function of its state. The change of "Vis vita" during motion measures the work performed.

The teaching of mechanical energy was completed at the beginning of the nineteenth century by the English scientist Thomas Young. He was the first to suggest the use of the term "energy" instead of "Vis vita" as a measure of the ability of a moving body to perform work. Elastic bodies preserve their energy after collision.

The principle of energy conservation was thus formulated only for mechanical phenomena. *Heat* was considered as a substance. However, James Watt had discovered that heat could actually perform work (heat engine), and this prompted the inclusion of thermal phenomena into the principle of energy conservation. This was done by the brilliant young French scientist

Sadi Carnot, who was the first to estimate the numerical value of the mechanical equivalent of heat (1824). In his pioneer work, Carnot considered heat as a heavy liquid, and thus the transfer of heat from a hot body to a cold one was equivalent to a drop in weight in the field of gravity. The temperature difference, ΔT, here played the role of height, and the quantity of heat, Q, played the role of weight.

The "substantial" theory of heat was generally accepted by the scientific community (remarkable exceptions were Young, Frenel, and Ampère) up to the middle of the nineteenth century. Very soon after his first work Carnot also disregarded this theory. In his great book, *Reflexions sur la puissance motrice du feu, et sur les machines propres à developper cette puissance* Carnot stated that heat is a form of energy and can be transformed into other forms. This book was written in 1824, however, it was only published by Carnot's brother in 1872, 40 years after the author's death.

Further developments in energy teaching in the nineteenth century were connected with Mayer, Joule, Helmholtz, and Clausius. Helmholtz was the first to formulate a general principle of energy conservation. According to Helmholtz, all types of energy can be subdivided into two main kinds: the energy of motion (now kinetic energy) and the energy of tension (now potential energy). In a closed system the sum of both forms of energy is a constant.

Clausius introduced the name "thermodynamics," as well as the notion of internal energy which is a "function of point," i.e., an exact differential. Having a property of potential, its change does not depend on the pathway of a system transition between two states. In other words, internal energy is a function of the state of the system. This is not true for heat and mechanical work. Discussing the details of the functioning of thermal machines (the Carnot cycle), Clausius discovered and introduced to science a new function of state—entropy. As a matter of fact, he had completed the construction of contemporary phenomenological thermodynamics. Until the present day, this science has served as the foundation of chemical thermodynamics and chemical kinetics. In turn, these fields of thermodynamics are used as generally accepted tools for the description and understanding of the energetics of intracellular processes.

Returning to the discussion between Leibnitz and Newton, and further extrapolating it to biochemistry and biophysics, we can say that Leibnitz won his battle against Newton. "Energy" is the most frequently used word in scientific books and articles. Almost nobody continues to use force to describe chemical reactions. This became an obstacle in the development of science.

2.2. Kinetics and Thermodynamics of Chemical Reactions

In this section we will briefly outline some of the conventional approaches of classical physical chemistry to the thermodynamics and kinetics of chemical reactions. It is necessary to clarify what the real assumptions underlying these approaches are.

Let us consider the scheme of a chemical reaction

$$\alpha_1 A_1 + \alpha_2 A_2 + \cdots + \alpha_n A_n \underset{k_b}{\overset{k_f}{\rightleftarrows}} \beta_1 B_1 + \beta_2 B_2 + \cdots + \beta_m B_m. \qquad (2.1)$$

Here, A_i $(i = 1, 2, \ldots, n)$ denote the chemical substances in the reaction volume before their chemical transformation (initial products), α_i are their stoichiometric coefficients, B_j $(j = 1, 2, \ldots, m)$ are the chemical substances after chemical transformation (final products), and β_j are their stoichiometric coefficients. The words "initial" and "final" are rather relative: the reaction is assumed to be reversible. For the sake of convenience the left side substances are always called "initial products," and the right side substances are called "final products."

Stoichiometric coefficients indicate the relation between the numbers of appearing and disappearing molecules of different chemical species. For instance, in the reaction

$$2H_2 + O_2 \rightleftarrows 2H_2O$$

the appearance (or disappearance) of two water molecules is accompanied without fail by the disappearance (or appearance) of two hydrogen and one oxygen molecules. If we know the scheme of the reaction and the corresponding stoichiometric coefficients, then the extent of the advance of a chemical reaction and its rate can be estimated by measuring the appearance or disappearance of any participant of the reaction. It is convenient to use the so-called "chemical variable," ξ, for the quantitative description of reaction (2.1), which can be characterized by different stoichiometric coefficients. Parameter ξ was introduced by De Donder [4] as the measure of the advancement of a chemical reaction. The variable ξ corresponds to the relative number of chemical transformations in the reaction. For example, in the course of the forward reaction (2.1), initiated in the vessel containing the unimolar concentrations of reagents A_i $(i = 1, 2, \ldots, n)$, the unit change of this parameter, $\xi = 1$, would correspond to the disappearance of α_1 moles of A_1, α_2 moles of $A_2, \ldots, \alpha_n$ moles of A_n, and the appearance of β_1 moles of B_1, β_2 moles of $B_2, \ldots, \beta_m$ moles of B_m, etc.

There are two main ideas in the foundation of classical chemical thermodynamics and in chemical kinetics: the notion of dynamic equilibrium and the law of mass action. Historically, formulating the mass action law Van't Hoff proposed that the reaction rate was determined by the concentrations of reacting molecules [5]. The elementary acts of chemical transformations in forward and backward reactions can proceed independently. According to the notion of dynamic equilibrium, the chemical equilibrium is established when the rates of forward and backward reactions become equal.

By definition, the rate of a chemical reaction, V, is determined by the number of chemical transformations in the unit of reaction volume in a unit of time, i.e., $V = d\xi/dt$. Mass-action law postulates that the reaction rate is proportional to the concentrations of reacting substances, $[A_i]$ or $[B_j]$, to the power of the corresponding stoichiometric coefficients, α_i or β_j. Thus, the rates of forward and backward reactions, V_f and V_b, will be determined by the following terms:

$$V_f = k_f [A_1]^{\alpha_1} [A_2]^{\alpha_2} \cdots [A_n]^{\alpha_n}, \qquad (2.2)$$

$$V_b = k_b [B_1]^{\beta_1} [B_2]^{\beta_2} \cdots [B_m]^{\beta_m}, \qquad (2.3)$$

where k_f and k_b are the rate constants of the corresponding reactions which can be determined as the rates of these reactions at the unit concentrations of all reagents. For the reactions in a gaseous phase, the partial pressures of reacting components have to be used instead of their concentrations.

The resulting rate of reaction (2.1) in the forward direction is $V = V_f - V_b$. Under the condition of chemical equilibrium $V_f - V_b = 0$, this leads to

$$\frac{[B_1]_{eq}^{\beta_1} [B_2]_{eq}^{\beta_2} \cdots [B_m]_{eq}^{\beta_m}}{[A_1]_{eq}^{\alpha_1} [A_2]_{eq}^{\alpha_2} \cdots [A_n]_{eq}^{\alpha_n}} = \frac{k_f}{k_b} = K_{eq}^*, \qquad (2.4)$$

where K_{eq}^* is the equilibrium constant. Such a "kinetic" approach to the introduction of the equilibrium constant, K_{eq}^*, does not imply the compulsory fulfillment of microscopic reversibility. This means that the pathways of forward and backward reactions can, in principle, be different. The equalizing of the rates of forward and backward reactions does not need the obligatory coincidence of the reaction pathways. It should be noted that (2.4), associating the equilibrium concentrations of the reagents, $[A_1]_{eq}, [A_2]_{eq}, \ldots, [A_n]_{eq}, [B_1]_{eq}, [B_2]_{eq}, \ldots, [B_m]_{eq}$, with the equilibrium constant K_{eq}^*, is usually called the law of mass action in the contemporary physical and chemical literature [6–8].

Postulating mass action law in the form of (2.2) and (2.3), and thus introducing "kinetically" the equilibrium constant K_{eq}^*, we also tacitly postulate the mechanism of reaction (2.1). As a matter of fact, (2.2) and (2.3) are valid, in general, only for the so-called elementary act of chemical reaction. The fulfillment of the mass action law, i.e., (2.2) and (2.3), *a priori* implies that the chemical transformations in both reactions, forward or backward, occur as the result of a "single" act of effective simultaneous collision of all reacting components. For reaction (2.1) the realization of forward or backward reactions in the chemical equation needs the simultaneous interaction of α_1 molecules of A_1, α_2 molecules of A_2, etc., or the simultaneous interaction of β_1 molecules of B_1, β_2 molecules of B_2, etc. This means that the rate constants k_f and k_b in (2.2) and (2.3) are determined by the cross section of scattering for the collisions of corresponding particles and the probabilities of the realization of chemical transformations in the course of these collisions. Equations (2.2) and (2.3) describe the average number of effective simultaneous collisions of reagents per unit time that cause the forward and backward chemical reactions. However, the real number of collisions fluctuates, deviating from the average value. Thus, for a detailed description of the reaction, which takes into account the fluctuations, one has to know the corresponding distribution functions for the reagents [7].

Rigorously speaking, the fulfillment of (2.2) and (2.3) also implies the fulfillment of certain physical conditions:

(i) The homogeneity of the reaction mixture; the concentration densities must be the same in each point of the reaction mixture. In practice, this condition is usually fulfilled under intensive stirring of the reaction mixtures, or can be provided by fast diffusion processes in the case of rather slow reactions.

(ii) The rate constants k_f and k_b are considered as the real constants which depend only on the temperature. The heat effects in chemical reactions might influence the constants k_f and k_b. To avoid complications in the calculations of reaction rates, the temperature of the reaction mixture must be kept fixed (by special technical precautions) in various points of the reaction volume.

(iii) Equations (2.2) and (2.3) imply that the frequencies of effective collisions are proportional to the concentrations of the reagents. From the physical point of view, this means that various components of the mixture (A_1, $A_2, \ldots, A_n, B_1, B_2, \ldots$, etc.) have one and the same distribution of their velocities. In order to provide the establishment of the Maxwell distribution of velocities, the ineffective elastic collisions of reacting components (i.e., those collisions which do not lead to chemical transformation) should be rather frequent. In this case, the frequency of molecule collisions would depend only on the concentration of the reagents but not on the peculiarities of the velocity distributions for different reagents.

(iv) Internal degrees of freedom for individual molecules must be in thermal equilibrium, which corresponds to the same temperature, T, as that characterizing the velocities of the molecules in the reaction medium. In this case, the frequency of effective collisions leading to chemical transformations would be independent of the internal degrees of freedom. As a matter of fact, this assumption excludes from consideration the long-living "excited" states of reacting molecules. However, as we will see later (Chapter 3), these long-living states can play a crucial role in the functioning of biological macromolecules.

Let us now consider the physical meaning of the "kinetically" introduced equilibrium constant, K_{eq}^* (2.4), and its connection with an equilibrium constant, K_{eq}, obtained within the frame of conventional equilibrium chemical thermodynamics. The equilibrium constant, K_{eq}^*, has been introduced above as the ratio of forward and backward rate constants and, in principle, can also be determined experimentally by measuring the corresponding equilibrium concentrations of the reagents, (2.4). In contrast to this "kinetic" approach, which tacitly implies the mechanism initiating the chemical transformation, the thermodynamic description of chemical reactions does not consider the reaction mechanism, nor does the pathway of reaching one or another state of the system. The classical thermodynamic approach to the description of chemical reactions, as well as to the "kinetic" approach, was also developed by Van't Hoff at the close of the nineteenth century [5]. The theoretical foundation for chemical thermodynamics lies in the fundamental works of the great American scientist J.W. Gibbs [9].

Let us now consider how to describe the chemical equilibrium in reaction (2.1) in the frame of equilibrium thermodynamics. By the way, we will answer the question as to: Whether or not the corresponding equilibrium constant would be identical to the "kinetic" equilibrium constant? According to equilibrium thermodynamics, the state of a system can be characterized by thermodynamic potentials which are the functions of volume (V), temperature (T), pressure (p), the number of particles (N), entropy (S), and other macroscopic parameters (x_i). Depending on the choice of macroscopic parameters, these potentials are: the internal energy $U = U(S, V, N, x_i)$, the Gibbs free energy $G = G(p, T, N, x_i)$, the Helmholtz free energy $F = F(V, T, N, x_i)$, and so on. The corresponding derivatives of thermodynamic potentials give other parameters characterizing the system (for details, see any textbook on physical chemistry, for example, [6]).

All biochemists traditionally use the Gibbs free energy, G, for a description of the energetics of biochemical processes. This function, as determined by the thermodynamic state of a system, is defined as

$$G = U + pV - TS = H - TS. \tag{2.5}$$

The condition of the thermodynamic equilibrium of a system corresponds to the minimum G value. The change in the Gibbs free energy is a measure of "useful work" that could be done in the course of a chemical reaction; the knowledge of the standard values of G changes in various biochemical reactions, ΔG^0, helps to evaluate the direction of the biochemical processes. The famous Gibbs equation, which combines the First and Second Laws of Thermodynamics, gives the relation (2.6) for an infinitesimal change in the G value, dG, occurring if a system performs work

$$dG = -S\, dT + V\, dp + \sum_i \mu_i\, dN_i, \tag{2.6}$$

where μ_i are the partial free energy of the ith component, and dN_i are the infinitesimal changes in the amount of the ith component of the reaction mixture. For the sake of simplicity, we omit here the term $\sum X_k\, dx_k$ associated with the changes in other macroscopic parameters x_k. Chemical potentials, μ_i, are the intensive factors defined as

$$\mu_i = \left(\frac{\partial G}{\partial N_i}\right)_{p, T, N_j(i \neq j)}. \tag{2.7}$$

For isothermal–isobaric conditions ($T = $ const. and $p = $ const., this is usually true for most biochemical systems), and other fixed macroscopic parameters ($x_j = $ const.), but with variable composition of the reaction mixture, we can write

$$(dG)_{p, T} = \sum_i \mu_i\, dN_i. \tag{2.8}$$

This value is equal to the so-called "chemical work" performed as the result of a chemical reaction. The tendency of any spontaneous chemical process will be determined by the fulfillment of the following condition:

$$(dG)_{p,T,x_j} \leq 0. \tag{2.9}$$

For the chemical reaction with known stoichiometric coefficients, the change in the amount of any reacting component, dN_i, can be given through the change in the degree of reaction advancement, $d\xi$. For the $d\xi$ acts of chemical transformation in reaction (2.1) we have the decrease in the numbers of the A_i molecule and an increase in the B_j molecules, which gives

$$dN_{A_i} = -\alpha_i \, d\xi; \qquad dN_{B_j} = \beta_j \, d\xi. \tag{2.10}$$

Using these relations, from (2.8) we obtain

$$(dG)_{p,T,x_j} = [-\sum \alpha_i \mu_{A_i} + \sum \beta_j \mu_{B_j}] \, d\xi. \tag{2.11}$$

From (2.11) are can obtain the value of the so-called *chemical affinity*, $\mathscr{A}$,

$$\mathscr{A} = -\left(\frac{dG}{d\xi}\right)_{p,T} = \sum \alpha_i \mu_{A_i} - \sum \beta_j \mu_{B_j}. \tag{2.12}$$

The value of the chemical affinity $\mathscr{A}$ of the reaction mixture determines the direction of the chemical reaction: at $\mathscr{A} > 0$ reaction (2.1) tends to the right-hand direction, and at $\mathscr{A} < 0$ it tends to the left-hand direction. In the state of chemical equilibrium, $\mathscr{A} = 0$.

For the so-called "ideal" chemical systems, composed of k components (including the solvent molecules) with their concentrations $[N_i]$ ($i = 1, 2, \dots, l$), chemical potentials are expressed as $\mu_i = \mu_i^0 + RT \ln n_i$, where μ_i^0 is a "standard" chemical potential and $n_i = [N_i]/(\sum [N_j]), j = 1, 2, \dots, l$, is a molar ratio of the ith component (see, e.g., [6]). Thus, in the case of reaction (2.1), we obtain

$$\mathscr{A} = \mathscr{A}^0 + RT \ln(n_{A_1}^{\alpha_1} n_{A_2}^{\alpha_2} \cdots n_{A_n}^{\alpha_n} n_{B_1}^{-\beta_1} n_{B_2}^{-\beta_2} \cdots n_{B_m}^{-\beta_m}), \tag{2.13}$$

where $\mathscr{A}^0 = \{\sum \alpha_i \mu_{A_i}^0 - \sum \beta_j \mu_{B_j}^0\}$ is a so-called "standard" affinity.

For rather diluted solutions, i.e., when the molar concentrations of all components, $[N_i]$, are much less than the molar concentration of the solvent, we can express $\mathscr{A}$ through the ratio of the molar concentrations of the reagents

$$\mathscr{A} = \mathscr{A}^0 + RT \ln([A_1]^{\alpha_1} \cdots [A_n]^{\alpha_n} [B_1]^{-\beta_1} \cdots [B_m]^{-\beta_m}). \tag{2.14}$$

By denoting $\mathscr{A}^0 = RT * \ln K_{eq}$, we introduce the equilibrium constant K_{eq}. For the equilibrium concentrations of reagents we obtain (2.15) which coincides with (2.4) for the equilibrium constant, K_{eq}^*, introduced above from the kinetic consideration

$$K_{eq} = \frac{[B_1]_{eq}^{\beta_1} [B_2]_{eq}^{\beta_2} \cdots [B_m]_{eq}^{\beta_m}}{[A_1]_{eq}^{\alpha_1} [A_2]_{eq}^{\alpha_2} \cdots [A_n]_{eq}^{\alpha_n}}. \tag{2.15}$$

Thus, for "ideal" systems, both the constants of chemical equilibrium, "thermodynamic" and "kinetic," have one and the same expression. Let us emphasize again that everything stated above is valid only for the elementary act of a chemical reaction. If reaction (2.1) is not an elementary act, but actually represents a scheme of sequential or parallel elementary acts, the kinetically and thermodynamically determined "equilibrium constants" may differ substantially.

Traditionally, in chemical and biochemical literature, the chemical affinity, $\mathscr{A} = -(dG/d\xi)_{p,T,x_j}$, and standard affinity, $\mathscr{A}^0 = RT * \ln K_{eq}$, are usually designated, correspondingly, as a free-energy change $-\Delta G$ and a standard free-energy change, $-\Delta G^0$. To avoid any misinterpretation, we have to make two remarks (see also [8]):

(i) ΔG and ΔG^0, which are usually expressed in energy units (Joules, calories) per mole, must be related to a generalized chemical coordinate, i.e., to the degree of advancement ξ, but not to the concentration of one or another reagent.

(ii) Conventional designations, "ΔG" or "ΔG^0" instead of $-\mathscr{A}$ or $-\mathscr{A}^0$, must not lead to the misunderstanding that "ΔG" or "ΔG^0" are referred to the total energy change in the course of a chemical reaction, from the initial state (determined by the concentrations $[A_1], [A_2], \ldots, [A_n]$, $[B_1], [B_2], \ldots$, etc.) to the final state of chemical equilibrium ($[A_1]_{eq}$, $[A_2]_{eq}, \ldots, [A_n]_{eq}, [B_1]_{eq}, [B_2]_{eq}, \ldots$, etc.).

Figure 2.1 demonstrates the traditional diagram for energy changes in the course of a chemical reaction. The abscissa axis represents the so-called "reaction coordinate," R, i.e., the generalized relative positions of all atoms of the reacting compounds and their change during transition from the left-hand to the right-hand state of the process (2.1). The ordinate axis is the total

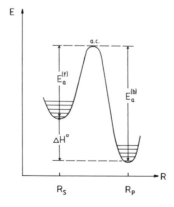

Fig. 2.1. Diagram illustrating the dependence of the system (substrate–product) energy, E, on the generalized reaction coordinate, R.

energy of the chemical system. Parameters presented in the scheme in Fig. 2.1, ΔH^0, $E_a^{(f)}$, and $E_a^{(b)}$, are usually employed for the description of chemical reactions in the frame of conventional chemical thermodynamics. Parameter ΔH^0 is a standard change in the enthalpy. The value ΔH^0 equals the "reaction heat," which is the total energy change caused by the chemical transformations in the system. "Kinetic" parameters $E_a^{(f)}$ and $E_a^{(b)}$ are the so-called "activation energies." Within the frame of the theory of the activated complex [10, 11], these parameters determine the rates of chemical transformation in forward and backward directions. We have to emphasize that, according to quantum mechanics, parameters ΔH^0, $E_a^{(f)}$, and $E_a^{(b)}$ should be counted off from the discrete energy levels but not from the bottom of the energy wells. In both wells, for initial and final products, the lowest levels are the so-called zero-point energy levels which correspond to the temperature $T = 0\ K$. Also, for an adequate thermodynamic description of the chemical reaction we usually need to know the standard change in entropy, ΔS^0. *The entropy change in the course of the chemical reaction would depend on the reaction pathway*, so it would be erroneous to plot the entropy value in this kind of energy scheme as demonstrated in Fig. 2.1.

The main equation of chemical thermodynamics connecting the parameters of a chemical reaction, ΔH^0 and ΔS^0, with the equilibrium constant, K_{eq}, is

$$\exp(\Delta S^0/R)\exp(-\Delta H^0/RT) = K_{eq}, \tag{2.16}$$

where R is a gas constant, and T is the temperature in degrees Kelvin. This equation is equivalent to (2.17)

$$\Delta G^0 = -RT \ln K_{eq}. \tag{2.17}$$

Here $\Delta G^0 = \Delta H^0 - T\Delta S^0$ represents a "standard" change in the Gibbs free energy in the course of a realized chemical reaction. We have seen above that in the case of so-called "ideal systems" the equilibrium constant, K_{eq}, coincides with the "kinetically" introduced equilibrium constant, K_{eq}^*, and can easily be expressed as the ratio of the corresponding equilibrium concentrations of reagents to the power of their stoichiometric coefficients, i.e., as the left side of (2.4). In principle, this allows us to determine experimentally the equilibrium constant, K_{eq}, by measuring the equilibrium concentrations of the reagents.

Having measured the temperature dependence of the equilibrium constant, K_{eq}, we can determine the ΔH^0 value using the Van't Hoff equation

$$d(\ln K_{eq})/d(1/T) = -\Delta H^0/R, \tag{2.18}$$

and then calculate ΔS^0 according to (2.16) and (2.17). The overwhelming majority of literature data, concerning the thermodynamic parameters of biochemical reactions, were obtained with the help of this Van't Hoff approach. Plotting the logarithm of the equilibrium constant K_{eq} versus $1/T$, experimentally estimated at various temperatures, the ΔH^0 values were estimated and then the ΔS^0 values were calculated. This widely used approach tacitly assumes that the parameters ΔH^0 and ΔS^0 are temperature-independent. Obviously, this is not a common case. It is possible, however, to indicate the condition under which (2.18) is also valid for the temperature-dependent parameters ΔH^0 and ΔS^0 [3]. Evidently, this is true if the following relation is fulfilled:

$$\partial(\Delta H^0)/\partial T = T\,\partial(\Delta S^0)/\partial T. \tag{2.19}$$

Formula (2.19) holds true, for example, for Kirchhoff's case:

$$\begin{aligned}\Delta H^0 &= (\Delta H^0)^* + \sum C_p^i T, \\ \Delta S^0 &= (\Delta S^0)^* + \sum C_p^i \ln T,\end{aligned} \tag{2.20}$$

where C_p^i is the heat capacity of the ith reactant at constant pressure. In the equation, the C_p^i of the final products of the reaction enter with sign " $+$," and those of the initial products enter with sign " $-$."

In the frame of the conventional thermodynamic approach, the change in ΔH^0 and ΔS^0 with the temperature is caused exclusively by the change in the Boltzmann distribution of reacting molecules, while the profiles of the potential wells (Fig. 2.1) remain temperature-independent. As we will see below, the latter condition could be violated in many biochemical reactions.

There exists a rather simple method for testing, as to whether or not the ΔH^0 determination with the help of the Van't Hoff equation is valid (2.18). The heat effect of the chemical reaction can be measured directly using a calorimetric technique, or calculated from an adequate Hess cycle using the available thermodynamic data. However, there are no methods for the direct measurement of energy parameters that determine the rate of a chemical reaction. According to the famous Arrhenius equation, the rate constant of a chemical reaction is:

$$k = A\,\exp(-E_a/RT). \tag{2.21}$$

Here the E_a value is the activation energy (see Fig. 2.1) that corresponds to the reaction advance per mole of the chemical variable ξ. The pre-exponential factor A is assumed to be temperature-independent; in this case

$$d(\ln k)/d(1/T) = -E_a/R. \tag{2.22}$$

Therefore, E_a can easily be calculated, plotting the experimentally measured $\ln k$ versus $1/T$. According to the simple collision theory, an act of chemical reaction can only occur if colliding molecules have the kinetic energy which exceeds the activation barrier, E_a. The "frequency" factor, A, is the number of collisions of reacting molecules per unit time. The exponential term in (2.21) determines a portion of those collisions which can lead to the chemical transformation. Note that (2.21) postulates the fulfillment of the Boltzmann equilibrium distribution of molecular energies in the reaction mixture.

A more sophisticated approach has been developed by Eyring in the theory of an activated complex [10, 11]. The generalized reaction coordinate, the R value in Fig. 2.1, must be calculated using the methods of quantum mechanics [12]. The coordinate R symbolizes the path between the initial and final molecular configurations which should be passed in order to overcome the lowest possible potential barrier. The maximum of this barrier (point "a.c." in Fig. 2.1) is a saddle point: it represents the potential energy maximum along the reaction coordinate and the potential energy minimum along other directions on the multidimensional map of the potential energy of a system. The statistical treatment used in this approach to the calculation of the rate constant seems to be rather contradictory. It has been assumed that there exists thermodynamic equilibrium between the initial state and the "activated complex" which is also maintained in the course of a chemical transformation. A somewhat simplified form of the Eyring equation for the rate constant can be written as follows:

$$k = (k_B T/h) \exp(S_a/R) \exp(-H_a/RT). \tag{2.23}$$

Here, k_B is the Boltzmann constant, h is the Planck constant, S_a is the entropy of activation, and H_a is the enthalpy of activation.

The activation enthalpy, H_a, in (2.23) plays the role of activation energy, E_a, in the Arrhenius equations (2.21) and (2.22). In a number of textbooks, dealing with the transition-state theory of chemical reaction kinetics, we can find the formula

$$k = ek_B(T/h) \exp(S_a/R) \exp(-E_a/RT). \tag{2.24}$$

Formula (2.23) transforms into (2.24) if the "experimental activation energy," E_a, is used in place of the theoretical "enthalpy of activation," H_a, [10, 11]. Let us repeat the traditional derivation of formula (2.24). E_a is determined from the temperature dependence of the reaction rate constant using the Arrhenius equation (2.22). On the other hand, it follows from (2.23) that:

$$d(\ln k)/d(1/T) = -T - H_a/R. \tag{2.25}$$

Therefore,

$$H_a = E_a - RT. \tag{2.26}$$

Substituting (2.26) into (2.23) we obtain (2.24).

As a matter of fact, this reasoning is wrong. In (2.23) the entropy and enthalpy of the activation are constants, or temperature-dependent but interconnected by the obvious equation (2.27) analogous to (2.19)

$$\partial E_a/\partial T = T \, \partial S_a/\partial T. \tag{2.27}$$

In principle, the values of these constants for any given system can be calculated. In the empirical equation (2.21) E_a is constant by definition.

The curves described by (2.21) and (2.23) cannot coincide at any values of the parameters. Equalization of (2.22) and (2.25) has only the following obvious meaning. For given values of two mutually independent parameters, H_a and E_a, we can find the temperature for which the slopes of the curves determined by (2.22) and (2.25) in the coordinates $\ln k$ versus $1/T$ are equal. When an experimentalist applies the activated complex theory calculating S_a, and substituting E_a for H_a in (2.23), he tacitly assumes that the activation barrier is equal to the enthalpy of the transition from the initial products to the activated complex, and that the temperature dependence of the frequency factor is negligible. There is no other meaning for the use of E_a in the formulas of the activated state theory.

As was pointed out above, the state of the activated complex is not a real molecule, and thus cannot be fixed and "measured" directly. Its life-time bears the property of infinitesimal quantity. Thus, the rate of the reaction is just the number of molecules undergoing the act of chemical transformation per unit time, i.e., the number of molecules passing through the state of the activated complex (point "a.c." in Fig. 2.1) per unit time. The theory of the activated complex assumes this passage in itself to be infinitely fast. This means that the reaction rate is determined by the number of chemical transformations per time unit but not by the duration of an individual act.

The treatments of chemical kinetics within the frame of the Arrhenius and the Eyring approaches were essentially based on the postulates of classical statistical equilibrium thermodynamics. It was assumed that a chemical system must pass through the sequence of equilibrium states. The principle of microscopic reversibility holds true all the way from the initial to final products. This implies that the pathways of the forward and backward reactions coincide. We have mentioned above that there exists a method of verification of the validity of thermodynamic equations used for the determination of the reaction enthalpy change. The heat production, ΔH, can be measured directly using a calorimetric technique. This cannot be done for the activation energy, E_a. It is necessary, therefore, to scrutinize the applicability of the conventional approaches of physical chemistry for a description of biochemical processes.

The kinetic and thermodynamic approaches briefly reviewed in this chapter were developed, as we have mentioned above, at the end of the nineteenth century for the chemical reactions of low-molecular compounds in a gaseous phase or dilute solutions. Are they applicable to biochemical reactions or not? *This is the question.* Obviously, the mechanisms of many complex chemical reactions or biochemical processes, in which biopolymers (proteins and nucleic acids) take part, are much more complex than in the case of low-molecular compounds. So, the main postulates lying in the foundation of conventional equilibrium chemical thermodynamics might be faulty (see Chapters 3 and 4).

2.3. Applicability of Equilibrium and Nonequilibrium Thermodynamics to Biological Systems and Processes

About twenty years ago a series of articles was published in the scientific literature, in which authors strongly confronted the notion of contemporary bioenergetics, especially with the concept of ATP as a universal energy keeper (for references, see [3]). These papers were criticized by many authors, and are almost completely forgotten nowadays. However, certain questions raised in these articles are not without interest even today. Among them is the question of the applicability of the laws and equations of thermodynamics to biochemical systems which, as a rule, are far removed from thermodynamic equilibrium. In [13], B.E.C. Banks and C.A. Vernon considered the process of glucose oxidation as an appropriate example for discussing the problem. The "equilibrium constant" of the overall process

$$C_6H_{12}O_6 + 6O_2 = 6CO_2 + 6H_2O + (2.7 \times 10^3 \text{ kJ/mole})$$

is about $10^{500} \, M^{-1}$ (a practically constant concentration of water is incorporated into this value). It is quite clear that the equilibrium can be reached only after a practically complete oxidation of glucose. Therefore, this process proceeds within the cell under conditions far removed from equilibrium. From the formal point of view, we can say that it would be incorrect (or even meaningless) to use the equilibrium thermodynamics approach.

There are plenty of examples of this nature. In any galvanic cell the concentrations of ions undergoing electrode reactions within half-cells are far removed from the equilibrium state of the overall chemical process. In the Daniell cell, the overall reaction

$$Zn + Cu^{2+} = Zn^{2+} + Cu \tag{2.28}$$

has equilibrium constant $K_{eq} \cong 2 * 10^{18}$. Thus, true equilibrium would be established after practically completely dissolving a metallic zinc electrode and precipitating copper ions on a copper electrode in their metallic form. It is common knowledge, however, that the properties of a galvanic cell can be adequately described by the formulas of equilibrium thermodynamics.

In both these examples, glucose oxidation and the Daniell cell, the systems are in the nonequilibrium states for kinetic reasons: the concentrations of glucose, in the first example, Zn^{2+} and Cu^{2+}, in the second one, change extremely slowly as compared with the time required to reach a real state of equilibrium. We can consider the concentrations of reacting substances (glucose, Zn^{2+} and Cu^{2+}) practically constant at any reasonable time interval. Therefore, both processes can be treated as proceeding reversibly at fixed values of the system parameters.

We have to distinguish such *kinetically nonequilibrium systems* from *open systems*, considered within the frame of the Onsager–Prigogine formalism of conventional nonequilibrium thermodynamics. The latter approach implies that the nonequilibrium state (stationary or not) is supported by inputting energy or material from the outside. Otherwise, in the former case, the nonequilibrium state is simply a consequence of the fact that such "kinetically nonequilibrium systems" do not have enough time to relax to the equilibrium state. We will see in the following chapters that the existence of slowly relaxing parameters, or so-called slowly relaxing degrees of freedom, is a compulsory requirement for using the system transient along these degrees of freedom in order to perform external work.

There is also another important problem concerning the nature of kinetically nonequilibrium states of chemical systems. This problem can be formulated in the following way. What are we dealing with, a nonequilibrium mixture of equilibrium molecules or nonequilibrium molecules? In any chemical process there appear molecules in nonequilibrium states. For low-molecular compounds, the electronic and vibrational relaxation after the elementary chemical act takes little time (as a rule, less than 10^{-11}–10^{-12} s). Therefore, there are relaxed atoms, ions, free radicals, and molecules, i.e., the particles in their equilibrium states, that take part in the subsequent chemical acts of chemical transformations. If a system consisting of low-molecular compounds is removed from the state of chemical equilibrium, then, as a rule, we can speak of a nonequilibrium ensemble of equilibrium molecules.

The difference between the chemical behavior of low-molecular compounds and highly ordered macromolecular systems (proteins, and nucleic acids) is a consequence of the fact that in the latter case the relaxation to the equilibrium state after local perturbation of each individual macromolecule can take a rather long time. Thus, a macromolecule ensemble can be considered as the mixture of nonequilibrium molecules.

There are many biochemical processes where we could expect the appearance of nonequilibrium states of macromolecules. The most important of them are:

(i) *Enzyme catalysis.* The formation of a substrate–enzyme complex can be regarded as the local chemical perturbation leading to rather slow conformational relaxation of this complex. The substrate–product transformation takes place in the course of this relaxation.
(ii) *Muscle contraction.* The local chemical acts (ion adsorption, ATP binding and hydrolysis) are accompanied by relatively slow conformational transients in contractible proteins.
(iii) *Intracellular ATP synthesis* in the energy-transducing membranes of chloroplasts, chromatophores, and mitochondria.
(iv) *The active transport of ions* across biological membranes leading to the generation of transmembrane electric potentials and the difference in the concentrations of ions. This transport is energetically ensured by the enzymatic hydrolysis of ATP.

All these processes may proceed under conditions excluding the fulfillment of certain postulates of conventional chemical thermodynamics and chemical kinetics. Let us analyze the possible consequences of using the formulas derived for low-molecular compounds.

For complex molecules (e.g., proteins) the temperature change may lead not only to changes in the Boltzmann distribution of molecules over the energy levels, but also to changes in the protein structure. Conformational changes could influence the profiles of potential wells characterizing the processes (Fig. 2.1). This means that formulas (2.16) and (2.18) may not hold true, and thus the use of the Van't Hoff equation in thermodynamics, as well as the Arrhenius (or Eyring) equation in kinetics, may lead to erroneous results. For instance, slight changes in the protein configuration with temperature variations could cause considerable changes in the activation barrier for the chemical transformation of a substrate attached to the enzyme (due to the strong dependence of chemical bond energy on the interatomic distance) without any relevant changes in the true activation entropy, S_a. Such an assumption was made as early as 1960 [14], in order to explain certain peculiarities in the kinetics of a free radical recombination in solids. In the narrow temperature range (as usual for the study of enzymatic reactions) any rather weak dependence of the activation energy, E_a, may be approximated by the linear law $E_a = E_a^0 + bT$, E_a where E_a^0 and b are constants. Introducing E_a into (2.21) or (2.23), we obtain

$$k = A \exp(-b/R) \exp(-E_a^0/RT), \tag{2.29}$$

$$k = (k_B T/h) \exp[(S_a - b)] \exp[-E_a^0/(RT)]. \tag{2.30}$$

In this case, the experimental data will formally satisfy the Arrhenius equation. However, in this case, we will measure not the true activation parameters E_a and S_a but the effective quantities, $E_a^0 = E_a - bT$ and $(S_a - b)$, that may be significantly different from the true values, E_a and S_a. It is clear that the E_a^0 value is in fact, the extrapolated value obtained from the real temperature dependence of the E_a value (Fig. 2.2). Even if E_a depends weakly on the temperature in certain intervals of temperatures, $T_1 - T_2$, the difference between the real and apparent values of the activation energy may be rather large.

The same reasoning can be used for the Van't Hoff equation (2.18). However, the correction for the ΔH^0 value, calculated from the temperature dependence of the equilibrium constant K_{eq}, can be checked using the calorimetric technique.

As we have emphasized above, in classical chemical kinetics the reaction rate is determined as the number of instantaneous elementary acts performed in a unit of time. However, if the conformational relaxation of a protein is included in the elementary chemical act, the latter cannot be considered as instantaneous. Therefore, we cannot exclude the possibility of those cases

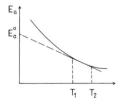

Fig. 2.2. Sketch of the possible temperature dependence of the activation energy, E_a.

when the rate of a chemical transformation is determined by the rate of relaxation of the protein structure to the new equilibrium state.

We have already stated in the Introduction that biopolymers are not only statistical but mechanical systems as well. This means that in many cases, when we are dealing with the transient (nonequilibrium) states of macromolecules, and not just with the quasi-equilibrium initial and final states of a system, the use of free energy becomes meaningless, and, thus should be abandoned. Mechanics uses only total energy, i.e., the Hamiltonian function. Such a "mechanical" approach to bioenergetic processes must include the notion of force.

Another peculiar property of biochemical processes is the small dimensions of their participants (macromolecules and macromolecular complexes, i.e., "molecular machines," closed vesicles, etc.). We will see that this also compels scientists to reconsider conventional approaches (see, for details, Section 3.3).

2.4. The Mechanisms of Energy Coupling in Chemical Reactions

Understanding the mechanisms of performing work from chemical reactions is deeply related to understanding the mechanisms of energy transfer between coupled energy-donating and energy-accepting chemical reactions. There are two extreme cases which can be called the mechanisms of indirect and direct coupling. *The mechanism of indirect coupling implies that there is no direct transfer of energy*, released in the course of an energy-donating process, to those components the chemical transformation of which needs energy. Energy exchange occurs via a thermostat. *The mechanism of indirect coupling has a statistical nature*: energy exchange occurs through thermal degrees of freedom distributed over the bulk phase of the reaction mixture. Free energy changes, in general, can have an entropic as well as an enthalpic nature, $\Delta G = \Delta H - T \Delta S$. As will be shown later, the enthalpy term, ΔH, cannot play a significant role in this type of coupling. In the case of indirect coupling it is the entropy term of free energy change (i.e., the concentration-dependent term, $- T \Delta S$) that determines the direction of chemical transformation in an energy-accepting reaction. The quantitative description of coupling processes is based on the mass action law, which determines the relationship between the reagents' concentrations in the volume of a common reservoir.

The mechanism of direct coupling implies that a significant part of the energy released in the course of an energy-donating reaction (enthalpy component, ΔH) is stored initially within a local domain (enzyme macromolecule or multienzyme complex), and can then be used to perform chemical work without energy spreading over thermal degrees of freedom in the environment. This type of coupling can be achieved by a single macromolecule working as a molecular machine. Being itself a statistical system, each individual macromolecule (molecular machine) can cyclically perform work practically independently of the states of other macromolecules. In this respect, *the behavior of molecular machines closely resembles that of mechanical systems* (for more details see Chapters 3 and 4).

2.4.1. Indirect Mechanism of Energy Coupling in Equilibrium (Quasi-Equilibrium) Homogeneous Mixtures of Chemical Reagents

Traditionally, chemical reactions in biochemistry are described by defining the concentrations of reacting components and chemical affinities that determine free energy changes in the course of chemical reactions. According to the mass action law, changes in the concentrations of some components affect those of other components by shifting the equilibrium in the reaction mixture. In this way, in principle, it is possible to realize indirect coupling between energy-donating and energy-accepting chemical reactions.

The general condition for coupling between energy-donating and energy-accepting processes (including those chemical processes which do not have common intermediates) can be formulated in the frame of the Onsager–Prigogine nonequilibrium thermodynamics [15–18]. Entropy production $\sigma_i = dS_i/dt = T^{-1}\mathscr{A}V \geq 0$, where V is the steady state rate of the chemical reaction and $\mathscr{A}$ is the corresponding chemical affinity. In a general case, with several chemically independent processes

$$\sigma_i = \sum_k X_k J_k, \tag{2.31}$$

where J_k is the rate of the kth irreversible process (chemical reaction, thermal flux, diffusion, and so on) in the system, X_k is the corresponding generalized force (chemical affinity, temperature gradient, gradient of the chemical potentials, and so on). Prigogine had emphasized [17] that the relationship (2.31) could be valid only in the vicinity of the system equilibrium state. Phenomenological thermodynamics of irreversible processes usually postulate linear relationships between the forces and fluxes,

$$J_k = \sum_m L_{km} X_m.$$

According to the conventional thermodynamic approach, changes in the value of free energy would determine the direction of the chemical transformation. Coupling two reactions becomes possible if a free energy decrease in one reaction ($\Delta G_1 < 0$ corresponding to the chemical affinity of this partial reaction $\mathscr{A}_1 > 0$) can compensate for a free energy increase in another reaction ($\Delta G_2 > 0$, $\mathscr{A}_2 < 0$), i.e., if $\mathscr{A}_1 + \mathscr{A}_2 > 0$. Increasing the free energy G_2 is equivalent to performing chemical work. This work can be done by shifting the chemical equilibrium in an energy-accepting reaction toward the formation of new molecules in a final product.

How can decreasing the free energy in the course of an energy-donating reaction ($\Delta G_1 < 0$) provide shifting the equilibrium of the energy-accepting reaction ($\Delta G_2 > 0$)? The elementary acts of energy-donating and energy-accepting reactions proceed independently. They, however, might be interconnected by two possible ways of indirect coupling. The first mechanism has a purely enthalpic nature, while the second one is purely entropic. Both mechanisms use the heat energy of the environment (heat energy spread over thermal degrees of freedom through the whole volume of the reaction mixture). In both cases, it is shifting the chemical equilibrium of the energy-accepting reaction that can be done either by changing the equilibrium constant due to a temperature increase ("*enthalpic*" indirect coupling) or due to the variations in the reagent's concentrations ("*entropic*" indirect coupling). We will see below that the efficiency of coupling chemical reactions through a thermal exchange of energy in the bulk phase of the reaction mixture cannot be practically efficient.

2.4.1.1. Enthalpic Mechanism of Indirect Coupling

We will start our analysis by evaluating the efficiency of the "enthalpic" mechanism of indirect coupling based on the dependence of the equilibrium constant on temperature. Let us consider, for the sake of simplicity, two monomolecular reactions which, in the general case, do not have common intermediates

$$A \underset{k_{-1}}{\overset{k_1}{\rightleftarrows}} B, \tag{2.32}$$

$$S \underset{k_{-2}}{\overset{k_2}{\rightleftarrows}} P. \tag{2.33}$$

We assume that reaction (2.32) is a heat-producing (exothermic) one, $(\partial H_1/\partial \xi_1)_{p, T} < 0$, while reaction (2.33) is a heat-consuming (endothermic) process, $(\partial H_2/\partial \xi_2)_{p, T} > 0$. According to the Van't Hoff equation,

$$(\partial H/\partial \xi)_{p, T} = RT^2 \frac{\partial}{\partial T} \ln K(p, T);$$

for an excoergic process (2.32) a temperature increase will lead to the equilibrium constant $K_1 = k_1/k_{-1}$ decrease, while for an endoergic reaction (2.33) the equilibrium constant $K_2 = k_2/k_{-2}$ will increase. Therefore, heating the reaction medium, produced by reaction (2.32), would lead to increasing K_2, thus shifting the reaction (2.33) equilibrium, and creating new molecules in the final product P. This mechanism of coupling does not compulsorily imply that both reactions should have common chemical intermediates. The exchange of energy (the enthalpy term of free energy change) occurs through the thermal degrees of freedom of a common reservoir. In real biochemical systems, however, this pure "enthalpic" mechanism of coupling cannot be practically realized. Most biochemical reactions proceed at an essentially constant temperature. This excludes the trivial effect of shifting the chemical equilibrium simply by heating (or cooling) the reaction mixture, that might be caused by heat production (or consumption) in the course of the energy-donating reaction. This conclusion can be illustrated by the following calculations.

We assume that reaction (2.32) is an exothermic one, i.e., each elementary act of the chemical transformation of the molecule A into molecule B is accompanied by a release of heat. For $\delta\xi_1$ elementary acts of reaction (2.32) heat production is $\delta Q_1 = -(\partial H_1/\partial\xi_1)_{p,T}\delta\xi_1$. Reaction (2.33) is assumed to be endothermic; and formation of $\delta\xi_2$ molecules P is associated with the heat absorption $\delta Q_2 = -(\partial H_2/\partial\xi_2)_{p,T}\delta\xi_2$. Since these reactions are chemically independent, the only way that reaction (2.32) could cause shifting of the equilibrium in reaction (2.33) is by heating the common reservoir containing the components of both reactions. For an adiabatic system, after $\delta\xi_1$ elementary acts of reaction (2.32) the temperature of the system will increase from T to T'; $\Delta T = T' - T = \delta Q_1/(c_p V)$, where c_p is the partial heat capacity (i.e., per one unit of volume) of the reaction mixture of volume V at constant pressure p. Increasing the temperature will lead to increasing the equilibrium constant from $K_2(p, T)$ to $K_2(p, T')$, thus shifting the equilibrium in reaction (2.33) to the formation of new P molecules. To evaluate the reaction yield, $\delta\xi_2$, we can use the Van't Hoff equation

$$\left(\frac{\partial H_2}{\partial\xi_2}\right)_{p,T} = RT^2 \frac{\partial}{\partial T}\ln K_2(p, T). \tag{2.34}$$

From this equation we obtain $K(p, T')/K(p, T) = e^\lambda$, where $\lambda = \lambda_0(\delta\xi_1/V)$ and

$$\lambda_0 = \frac{\Delta T}{RT^2}\left(\frac{\partial H_2}{\partial\xi_2}\right)_{p,T} = -\frac{(\partial H_1/\partial\xi_1)_{p,T}(\partial H_2/\partial\xi_2)_{p,T}}{Rc_p T^2}. \tag{2.35}$$

Let p, s and p', s' be equilibrium amounts of P and S molecules at temperatures T and T', respectively. Then $K_2(p, T) = p/s$ and $K_2(p, T') = p'/s'$. If $\delta\xi_2$ is a thermoinduced advance of reaction (2.33), then

$$\frac{K_2(p, T')}{K_2(p, T)} = \frac{1 + \delta\xi_2/p}{1 - \delta\xi_2/s}. \tag{2.36}$$

The relative increase in the concentration of the final product P from S (parameter γ_2) is

$$\gamma_2 = \frac{\delta\xi_2}{s} = \frac{e^\lambda - 1}{K_2 e^\lambda + 1}K_2. \tag{2.37}$$

If $\lambda \ll 1$, the coupling stoichiometry $\eta = \delta\xi_2/\delta\xi_1 \cong K_2\lambda_0 s/V$. Thermoinduced coupling could be efficient only in the case $K_2 e^\lambda \gg 1$, when $\gamma_2 \cong 1$ (in the limit $\lambda \to \infty$). Since reaction (2.33) is assumed to be an energy-accepting one, then $K_2 < 1$. This means that the high efficiency of coupling could be achieved only if $\lambda \gg 1$. We will see below that under any reasonable experimental conditions, even for an extremely low volume of the reaction mixture, V, it is impossible to reach conditions when $\lambda = \lambda_0(\delta\xi_1/V) \gg 1$.

Let us evaluate parameter γ_2 taking $(\delta H_1/\delta\xi_1)_{p,T} \cong -55$ kJ/mole ("standard" heat of the reaction $H^+ + OH^- = H_2O$) and $(\delta H_2 / \delta \xi_2)_{p,T} = 30$ kJ/mole ("standard" heat of the reaction of ATP formation from ADP and P_i). Also, let the standard affinity for reaction (2.33) be $\mathscr{A}_2^0 = RT \ln K_2 = -30$ kJ/mole. This value $\mathscr{A}_2^0$ corresponds to the value of the equilibrium constant $K_2 = 6 \cdot 10^{-6}$. Taking for partial heat capacity (per one litre of aqueous solution) $c_p \cong 4.18$ kJ/mole K, we obtain $\lambda_0 \cong 0.527\ M^{-1}$. Since c_p values for proteins are greater than for pure water, for protein solutions $\lambda_0(\text{protein}) < \lambda_0(H_2O)$. Under any reasonable conditions $\delta \xi_1 / V \ll 1\ M$; therefore, it is practically impossible to have $\lambda \gg 1$, as well as to obtain a significant yield of P molecules from the energy-accepting reaction (2.33). Thus, the "*enthalpic*" mechanism of indirect coupling appears to be unrealistic.

This mechanism is inefficient in a purely statistical system (homogeneous reaction mixture) because energy released is shared by a tremendous number of thermal degrees of freedom. For any reasonable conditions, heating the reaction mixture will be too meager to cause shifting of the equilibrium in an energy-accepting reaction. However, the realization of an "enthalpic" mechanism becomes possible in the case of coupling performed by "molecular machines." The operation of molecular machines implies that at least part of the energy released does not dissipate over thermal degrees of freedom in the surrounding medium, but is stored in the local domain of the macromolecule in the form of the excitation of one (or a few) selected degrees of freedom of a molecular machine. This "stored" energy can then be directly delivered to perform chemical work (for further details, see Chapters 3 and 4).

2.4.1.2. Entropic Mechanism of Indirect Coupling

As early as 1905, A.N. Shilov pointed out that the mechanism of indirect coupling can be realized if the participants of both reactions, energy-donating and energy-accepting, were connected through a common chemical intermediate [19]. Shifting the equilibrium in an energy-accepting reaction can be initiated, for example, by increasing the concentration of one of its reagents produced in the course of an energy-donating reaction. This mechanism has a purely statistical nature. Since this mechanism of energy coupling is driven by the changes in the concentration of certain participants, it is the entropic (concentrational) part of the free energy changes that is responsible for producing chemical work. Such an approach can adequately describe the behavior of certain kinds of coupled chemical reactions. Let us now consider this approach in more details.

The energy of substrates is initially conserved in the form of potential energy of the chemical bonds in certain molecules. According to the conventional approach, energy released in the course of an energy-donating process, being distributed over a great number of thermal degrees of freedom in a common reservoir, is not delivered directly for performing work or increasing the potential energy of other participants of the reaction mixture, which chemical transformation is associated with energy consumption. Under equilibrium (or quasi-equilibrium) conditions reactants are at local thermodynamic equilibrium with their environment. Elementary acts of energy-accepting and energy-donating processes can be separated in time and space. If the energy evolved is spread over a tremendous number of thermal degrees of freedom of a macroscopic reservoir, and thus cannot be used directly for performing chemical work, the question arises: How can the energy-donating reaction stimulate overcoming the activation barrier in the course of an energetically unprofitable chemical transformation?

First of all, it should be stressed that in the case of an entropic mechanism of indirect coupling both reactions, energy-accepting and energy-donating, can be endothermic as well as exothermic. The role of the reservoir for dumping energy at several steps of the exothermic processes serves the environment. In order to ensure coupling (i.e., to provide the conditions for performing the energy-accepting reaction with negative chemical affinity) the decisive role has only the sign of free energy changes (positive affinity in an energy-donating reaction). The total energy of the system can be either decreased or increased, depending on the nature of the particular elementary act of an energy-accepting reaction. The energy to perform a chemical transformation is picked out or delivered to the environment (thermostat). Concentrational changes lead only to changing the probability of this elementary act. Thus, the environment plays the role of a thermostat which supplies energy for overcoming the potential barrier of chemical transformations. Elementary acts of these energy-accepting steps are associated with the selection of "hot" particles from the normally distributed reagents in the reaction mixture. The higher the activation barrier, E_a, the larger should be the energy of the reacting particle. With increasing temperature the probability of obtaining appropriate "hot" particles from the fluctuating medium becomes higher and, therefore, the reaction rate increases.

As we have mentioned above, indirect coupling becomes possible if both reactions have a common intermediate which is a product of the first reaction and a substrate for the second one. Increasing the total number of intermediates, N, concomitantly increases the number of "hot" particles. Thus, producing intermediate particles in the course of an energy-donating reaction increases the probability to "fish out" an appropriate "hot" particle which is able to overcome a potential barrier in the process of the chemical transformation. Facilitating the energy-accepting reaction, the increase in the concentration of intermediate particles shifts the equilibrium in favor of the formation of new molecules of the final product.

It is necessary to emphasize that using thermal degrees of freedom for driving energy-accepting steps of chemical transformations is not a violation of the Second Law. Quantitative analysis demonstrates that an entropic mechanism of indirect coupling can be efficient, in practice, only in open systems which are supported in their nonequilibrium states due to the steady exchange of substrate and product molecules with the environment. In steady state (or quasi-equilibrium) conditions, cooling of the system, that might take place due to the "selection" of "hot" particles, is readily compensated for by heat flow from the thermostat and heat release in the course of the energy-donating reaction. We want to emphasize again that this indirect mechanism of coupling has a statistical nature. Although each portion of energy for performing endothermic elementary acts is accepted from the thermal degrees of freedom of the environment, it is the entropic (concentrational) (but not enthalpic (heat)) component of the free energy change that ensures the performance of chemical work. As a matter of fact, increasing the concentration of the common intermediate, accompanied by the enhancement in the average number of "hot" particles, is equivalent to increasing the entropy of this component in the course of the energy-donating reaction.

Thus, in the frame of classical chemical thermodynamics and kinetics, there is no formal restrictions on performing chemical work using an indirect mechanism of coupling. However, quantitative analysis demonstrates that in a system functioning cyclically under equilibrium (or quasi-equilibrium) conditions the efficiency of this mechanism cannot be high. High stoichiometric ratios are possible only in the case of open system functioning under non-equilibrium conditions.

Let us consider the question of the efficiency of indirect coupling. We introduce two possible measures of indirect energy coupling. The first is determined as the ratio of the number of resulting elementary acts of an energy-accepting process, n_a, to the number of elementary acts of a coupled energy-donating process, n_d, i.e., $\gamma = n_a/n_d$. This measure can be used regardless of the method of initiating the energy-donating process (the addition of a substrate, the variation of temperature, the addition of a catalyst to the system that was initially kept in a nonequilibrium state). This measure will be further called *the efficiency of coupling*. The second measure is the ratio of the number of newly formed molecules of the product, n_a, to the number of added molecules of the substrate of the energy-donating reaction, n_{ad}, i.e., $\eta = n_a/n_{ad}$. This measure will be called *the stoichiometric ratio* or *the conversion factor*.

Let us first consider an isolated reactor: the mixture of A, B, and C molecules that are undergoing monomolecular reversible transformations

$$A \underset{k_{-1}}{\overset{k_1}{\rightleftarrows}} B, \tag{2.38}$$

$$B \underset{k_{-2}}{\overset{k_2}{\rightleftarrows}} C \tag{2.39}$$

Here, k_1, k_{-1}, k_2, and k_{-2} are the corresponding rate constants for the forward and backward reactions, and $K_1 = k_1/k_{-1}$ and $K_2 = k_2/k_{-2}$ are the corresponding equilibrium constants. Compound B is a common reagent for both reactions, being a product of reaction (2.38) and a substrate for reaction (2.39). Producing new molecules of C in reaction (2.39) is an unfavorable process, but it becomes possible by increasing the concentration of B in the course of the energetically profitable reaction (2.38). In a closed system the only reasonable measure of coupling efficiency is the first one (parameter γ).

Let us consider the relaxation of a system (with initial concentrations of the reagents a, b, and c) to chemical equilibrium. For certainty, we assume that the energy-accepting reaction $(B \Leftrightarrow C)$ was initially at equilibrium, i.e., $c/b = K_2$, while the energy-donating one $(A \Leftrightarrow B)$ was out of equilibrium. After reaching the equilibrium state the concentrations of the reagents will obey the equations $(b + n_d - n_a)/(a - n_d) = K_1$ and $(c + n_a)/(b + n_d - n_a) = K_2$. From these relations we finally obtain

$$\gamma = n_a/n_d = K_2/(1 + K_2) < 1.$$

This means that the coupling efficiency could be high only if $K_2 \gg 1$, i.e., for a thermodynamically favorable reaction. Meanwhile, for any energy-accepting reaction $(K_2 \ll 1)$ high efficiency cannot be reached.

Let us now consider a similar system that is initially kept in a state of equilibrium. The equilibrium is then shifted by adding a certain amount, Δa, of the substrate A of the energy-donating reaction. In this case, we can determine both the parameters of coupling, γ and η. To evaluate the stoichiometry of coupling, we need to calculate the number of new C molecules, Δc, produced in response to addition into the reaction mixture of Δa molecules of A. In this way we can determine the conversion factor $\eta = \Delta c/\Delta a$. After addition of Δa molecules of A to the equilibrium mixture, the system will reach a new state of equilibrium with the reagents' concentrations a', b', and c'. Using the condition of material balance, $a + b + c + \Delta a = a' + b' + c'$, based on the reactions stoichiometry, we finally obtain the following expressions for the efficiency of the coupling and conversion factors:

$$\gamma = K_2/(1 + K_2), \quad \text{and} \quad \eta = K_1 K_2/(1 + K_1 + K_1 K_2) < K_2/(1 + K_2).$$

We can see that the relative yield of C molecules depends only on the thermodynamics of coupled reactions. It follows from these formulas that it is practically impossible to reach high conversion factor values if an energy-consuming reaction (2.39) is characterized by a negative value of the standard chemical affinity, $\mathscr{A}_2^0 = RT \ln K_2 < 0$. For example, if $\mathscr{A}_2^0 \cong -30$ kJ/mole (standard affinity for the reaction of ATP synthesis from ADP and P_i), then $K_2 \cong 6 * 10^{-6}$, and consequently $\eta \ll 1$. This result illustrates why the mechanism of indirect coupling of the two chemical reactions having a common intermediate cannot be efficient. Actually, according to the above-mentioned example, in order to form one molecule C we need to add more than 10^6 molecules B.

A similar conclusion leads to the consideration of an another type of reaction

$$A + B \underset{k_{-1}}{\overset{k_1}{\rightleftarrows}} C. \tag{2.40}$$

Is it possible to reach a high efficiency in producing compound C due to shifting equilibrium, simply by means of the addition of one of the reagents (e.g., compound B) standing on the left side of (2.40)? In the case of many important biochemical reactions the role of component B might serve, for instance, hydrogen ions, H^+. Under equilibrium conditions $c/ab = K = k_1/k_{-1}$, where a, b, and c are the concentrations of A, B, and C, respectively. After the addition of x molecules B to the equilibrium mixture of A, B, and C, the system will reach a new state of equilibrium with the concentrations of the reagents a', b', and c', where $c' = c + \xi$, $a' = a - \xi$, and $b' = b + x - \xi$. Parameter ξ equals the number of newly formed molecules C; in other words, parameter ξ is the advance of chemical reaction (2.40) initiated by the addition of B. Elementary calculations lead to the following term for the conversion factor $\eta = \xi/x$:

$$\eta = 0.5(1 + \chi/x) * \{1 - [1 - 4ax/(\chi + x)^2]^{1/2}\}, \qquad (2.41)$$

where $\chi = a + b + K^{-1}$. For comparatively small additions of compound B, when $x \ll \chi$, we obtain $\gamma \cong aK/(1 + (a + b)K)$. Reaching a high stoichiometry of coupling, $\gamma \cong 1$, is possible only if $aK \gg 1$. For any reasonable condition, the substrate A concentration is $a \ll 1$ M. Hence, the fulfillment of the condition $aK \gg 1$ is equivalent to the condition $K \gg 1$. For this reason, the high efficiency of the promoting reaction (2.40) by the addition of compound $B(\gamma \to 1)$ becomes possible only for the reaction characterized by a positive standard chemical affinity $\mathscr{A}^0 = RT \ln K > 0$, i.e., in the trivial case of the energetically favorable process of C formation.

If reaction (2.40) is an energy-accepting process, characterized by the negative value of standard chemical affinity, $\mathscr{A}^0 < 0$ (or $K < 1$), the high stoichiometric ratio γ of coupling cannot be achieved. For instance, taking the input parameters that are typical for the reaction of ATP formation from ADP and P_i, we obtain $\gamma \ll 1$. Actually, for $K \cong 6 * 10^{-6}$ M^{-1} (corresponding to $\mathscr{A}^0 \cong -30$ kJ/mole) and substrate A concentration $a \cong 10^{-3}$ M, formula (2.41) gives $\gamma \cong 10^{-9}$. This means that the addition of component B to the equilibrium reaction mixture cannot drive the formation of a sufficient amount of new C molecules. It is easy to demonstrate that the relative yield of C molecules, $\eta = \xi/x$, will diminish with every new portion of substrate B added to the equilibrium reaction mixture. The maximal stoichiometric ratio for coupling, $\eta = \xi/x$, can be reached only for small enough quantities of the added component B, reaching a maximum in the limit $x \to 0$. The absolute yield of C molecules, $\xi = \gamma x$, increases with increasing x. However, in order to reach a reasonably high absolute yield of product C (parameter ξ), we would need to add comparatively large amounts of B.

All the above-considered examples demonstrate that the mechanism of indirect coupling cannot provide the high yields of the product in energy-accepting reactions, if the substrate molecules are simply added to equilibrium homogeneous reaction mixtures. High stoichiometry of coupling might be reached only in the trivial case of the energetically profitable reaction (2.40), the equilibrium of which is significantly shifted to the right side. On the other hand, the most important energy-transducing processes in living cells are characterized by high values of coupling stoichiometry. For example, under steady state conditions the ATP synthesis in biomembranes, associated with the transmembrane transport of hydrogen ions across coupling membranes of chloroplasts and mitochondria, demonstrates the stoichiometric ratio $ATP/H^+ \cong \frac{1}{3}$, i.e., the passage of three protons through the ATP-synthase can provide formation of one ATP molecule (for details, see Chapter 5). Of course, in both examples considered above, the reason for the low efficiency of coupling is not simply a consequence of energetic difficulties. As was noted above, under quasi-equilibrium conditions the reactants are able to extract energy from the thermostat. What could be the reasons for the high stoichiometry of coupling in real biochemical systems? There might be certain "kinetic" factors:

(i) nonequilibrium (steady state) conditions for operating the system;
(ii) the presence of special "devices" (construction) that support nonequilibrium conditions (recurrent addition of substrate and synchronous withdrawal of the product from the reaction mixture); and
(iii) compartmentalization (or "channeling") of chemical reactions (for details, see Section 3.3).

The foregoing examples also demonstrate that producing noticeable amounts of the final product of the energy-accepting reaction can be achieved, only in those unrealistic cases when the concentration of at least one of the substrates in the reaction mixture is significantly and recurrently increased. Formally, we could overcome this difficulty assuming, for instance, that the high extent of concentrating the reactants can be achieved by injecting a few (or even single) particles into a microscopic compartment or chemical "reactor" of very small volume. This might be one of the reasons why the compartmentalization of the reagents within relatively small subcellular compartments ensure the high stoichiometric ratios in coupling biochemical processes. Concerning the effects of the reagents' compartmentalization (or their direct "channeling" into small compartments), we have to take into account the following difficulties that can arise when using a formal thermodynamic description:

(i) under certain conditions (in small systems) the law of mass action, used as the conventional principle for calculating reagent concentrations, may be violated (for more details, see Section 3.3);
(ii) based on the law of mass action, the conventional deterministic approach does not take into account the discrete nature of reacting particles; in certain cases, the discreteness can be a reason for enormously large fluctuations, and for a quantitative description of such systems we need to use a stochastic approach (for references, see [7, 20]);
(iii) misleading use of the reagents' concentrations averaged over the total volume of the inhomogeneous systems (for instance, the reagents' compartmentalization within local domains or the vesicles of heterogeneous populations); and
(iv) coupling of biochemical reactions is performed by the enzymes; otherwise, the behavior of macromolecules in certain important respects can-

not be described within the formalism of the conventional chemical thermodynamics elaborated for low-molecular compounds (for further details, see Chapter 4).

All conclusions concerning relatively low-coupling efficiency are valid, of course, if we consider either a thermodynamically isolated closed system, or an open system with the unidirectional input of the substrate but without the withdrawal of the final product. In the latter case the system cannot reach steady state. This becomes possible only in open systems. To arrange a cyclic regime of C formation in reaction (2.40), we should recurrently withdraw from the reaction mixture an adequate quantity of newly produced molecules C, and then add corresponding amounts of the substrate B at every new turn of the chemical cycle. The behavior of open systems will be considered in the next section.

The above-considered mechanism of indirect coupling also faces some other obvious difficulties in the explanation of elementary steps of several energy-accepting processes (active transport, muscle contraction, oxidative- and photophosphorylation). As McClare stated [21], there is no immediate influence by the environment on the ability of ATPsynthase (or ATPase) to perform formation or hydrolysis of ATP. Really, under steady state conditions ATPsynthase (or ATPase) acts cyclically, whereas the chemical composition of the environment on both sides of the coupling membrane remains practically constant. Also, recurrent contraction and relaxation of myosin cross-bridges in the muscle's filaments can proceed cyclically if there is a sufficient amount of ATP and Ca^{2+}, but in practice these steps are not directly induced by the immediate change in the ATP and Ca^{2+} content in the medium. These examples imply that cyclic functioning of a single enzyme molecule (e.g., ATPase) must be triggered by the intrinsic processes, but does not need the recurrent changes in the concentrations of low-molecular substrates or products in the bulk phase of the environment.

2.4.2. Entropic Mechanism of Coupling Chemical Reactions in Open Systems

If the steady state concentrations of the components are shifted, but not too far from their equilibrium values, the interconnection between the fluxes and chemical forces (chemical affinities, in our case) should satisfy the well-known linear relationships that are usually postulated in the linear thermodynamics of irreversible processes [15–18]. We do not consider here the phenomenological equations of nonequilibrium thermodynamics. For details the reader can refer to numerous excellent monographs and review articles devoted to the applications of nonequilibrium thermodynamics in the description of chemical reactions and biological processes (see, for instance, [22–30]). In many cases, the conventional phenomenological approaches of linear and nonlinear nonequilibrium thermodynamics appear to be useful tools for the

quantitative analysis of the relationships between the fluxes and the corresponding thermodynamic forces; the interaction between coupled fluxes in
membrane processes, the efficiency of producing chemical work, the stability
of steady states, and so on.

Being a useful theoretical instrument, the phenomenological approach,
however, cannot clarify the molecular mechanisms of chemical processes.
The famous Onsager reciprocal relationships were derived in their general
form from the principle of the local reversibility of physical processes. For
chemical reactions, some useful flux–force relationships can be derived if we
know the mechanism of the reactions considered. In this way, in principle, it
becomes possible to express the phenomenological Onsager proportionality
coefficients through the kinetic constants of elementary steps.

We want to illustrate here the behavior of the open systems using the
simplest examples of monomolecular chemical reactions. For quantitative
analysis, in this section we will consider thermodynamic and kinetic relations
for coupled monomolecular reactions ($A \Leftrightarrow B$ and $B \Leftrightarrow C$). Analysis of the
behavior of this reaction in an open system gives explicit and simple relationships, between the kinetic and thermodynamic characteristics of the reaction
mixture on one hand, and the flux through the system on the other. However,
for didactic reasons we start our analysis with a consideration of the simplest
monomolecular reaction, the transformation of S into P.

As noted above, a simple injection of ΔS molecules of S into the *equilibrium
mixture of S and P* will lead to producing ΔP molecules of P with the stoichiometry ratio $\eta = \Delta P / \Delta S = K/(1 + K)$, where $K = k_+/k_-$ is the equilibrium constant, and k_+ and k_- are the forward and backward rate constants
for the reaction $S \underset{k_-}{\overset{k_+}{\rightleftarrows}} P$. For any energy-accepting reaction with $K \ll 1$ the
relative yield of P will be very low. *A high stoichiometric ratio would be
established if the mixture of components S and P is led out of equilibrium due
to a steady exchange with the surroundings*, that includes not only the input of
the substrate S but also the output of the product P from the reaction mixture. Let us consider the kinetic behavior of such an open system symbolized
by the following scheme:

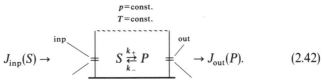

$$J_{\text{inp}}(S) \rightarrow \qquad S \underset{k_-}{\overset{k_+}{\rightleftarrows}} P \qquad \rightarrow J_{\text{out}}(P). \qquad (2.42)$$

In order to describe the transient behavior of the chemical system (at constant pressure, p, and temperature, T), we will consider the variable functions:
Gibbs free energy, $G(t)$, enthalpy, $H(t)$, and chemical potentials, $\mu_S = \mu_S^0 +
RT \ln S(t)$ and $\mu_P = \mu_P^0 + RT \ln P(t)$, where μ_S^0 and μ_P^0 are the standard
chemical potentials of S and P. Following Prigogine's line of argument [17,
27], we assume that these functions are determined by the same macroscopic
variables (concentrations, volume, temperature, pressure, and so on) which

determine the state of the equilibrium system. For any moment of time t, it is reasonable to determine $G(t)$ and $H(t)$, as related to one unit of volume, by the following functions

$$G(t) = S(t)[\mu_S^0 + RT \ln S(t)] + P(t)[\mu_P^0 + RT \ln P(t)], \qquad (2.43)$$

$$H(t) = S(t)\mu_S^0 + P(t)\mu_P^0. \qquad (2.44)$$

We realize that using these thermodynamic functions represents a formal but reasonable approach to the description of an open system. At any moment in time, the values of the thermodynamic functions, $G(t)$ and $H(t)$, determine the spontaneous relaxation of the reaction mixture to the equilibrium state that would take place after abrupt and simultaneous ceasing of the input and output fluxes.

Let us now consider the transient kinetics of the changes in the concentrations S and P, as well as the thermodynamic functions $G(t)$ and $H(t)$, that take place after simultaneous imposing (or ceasing) of the input and output fluxes. Let us assume that at the initial moment of time $t = 0$ the mixture of S and P had the component concentrations $S(0) = S_0$ and $P(0) = P_0$. In a general case, this initial state may be a nonequilibrium one. We assume that at the moment of time $t = 0$ two fluxes instantaneously appear, J_{inp} and J_{out}, corresponding to the input of S into the reaction mixture and the output of P from the mixture. The concentrations of S and P change with "switching on" the fluxes, symbolized on Scheme (2.42) by open channels "inp" and "out." The deterministic kinetic equations for the variable concentrations, $S(t)$ and $P(t)$, can be written as follows:

$$dS/dt = J_{inp} - k_+ S + k_- P, \qquad (2.45)$$

$$dP/dt = k_+ S - k_- P - J_{out}. \qquad (2.46)$$

The system can reach the steady state if the stoichiometric ratio of the input and output fluxes is $\eta = J_{out}/J_{inp} = 1$. The flux value, $J = J_{out} = J_{inp}$, is considered as the parameter of the model. The solutions of the equations (2.45)–(2.46) are following functions:

$$S(t) = S_0[1 + \alpha(t)] = S_0\left[1 + \frac{J_0}{k_0 S_0}\left(1 - e^{-k_0 t}\right)\right], \qquad (2.47)$$

$$P(t) = P_0[1 + \beta(t)] = P_0\left[1 - \frac{J_0}{k_0 P_0}\left(1 - e^{-k_0 t}\right)\right], \qquad (2.48)$$

where $J_0 = J - k_+ S_0 + k_- P_0$. The parameter $k_0 = k_+ + k_-$ determines the characteristic time, $\tau = k_0^{-1}$, of reaching the steady state concentrations $S^* = S_0 + J_0/k_0$ and $P^* = P_0 - J_0/k_0$. Physical meaning have only those values of parameter J that obey the condition $J < k_0(P_0 + S_0)$.

Let the fluxes be "switched on" at time $t = 0$. Enthalpy changes will be described by the following expression:

$$\Delta H(t) = H(t) - H(0) = \frac{J_0}{k_0}(1 - e^{-k_0 t})\,\mathscr{A}^0, \qquad (2.49)$$

where $\mathscr{A}^0 = (\mu_S^0 - \mu_P^0) = RT \ln K$ is the standard affinity, and $K = k_+/k_-$ is the equilibrium constant of the reaction $S \Leftrightarrow P$. The sign of the enthalpic effect depends on the $\mathscr{A}^0$ value: If $J_0 > 0$, then for an energy-accepting reaction ($\mathscr{A}^0 < 0$, or $K < 1$) the flux through the reaction mixture will be accompanied by heat consumption ($\Delta H > 0$), while for an energy-donating reaction ($\mathscr{A}^0 > 0$, or $K > 1$) the flux will lead to heat production ($\Delta H < 0$). In the course of reaching steady state, the enthalpy production fades exponentially

$$h = dH/dt = J_0 \mathscr{A}^0 e^{-k_0 t}. \qquad (2.50)$$

The transient kinetics of Gibbs free energy changes, $\Delta G(t) = G(t) - G(0)$, depends not only on the kinetic parameters but also on the initial chemical affinity, $\mathscr{A}(0) = \mathscr{A}_0$,

$$\Delta G = \frac{J_0}{k_0}(1 - e^{-k_0 t})\mathscr{A}_0 + RTS_0(1 + \alpha)\ln(1 + \alpha) + RTP_0(1 + \beta)\ln(1 + \beta).$$

The sign of the $G(t)$ changes can be determined by the sign of its derivative

$$dG/dt = J_0\left[\mathscr{A}_0 + RT \ln\left(\frac{1 + \alpha(t)}{1 + \beta(t)}\right)\right]e^{-k_0 t}, \qquad (2.51)$$

where

$$\alpha(t) = \frac{J_0}{k_0 S_0}(1 - e^{-k_0 t}) \qquad \text{and} \qquad \beta(t) = -\frac{J_0}{k_0 P_0}(1 - e^{-k_0 t}).$$

If, in the initial state, $\mathscr{A}(0) = \mathscr{A}_0 = 0$, after switching on the fluxes, the free energy of the reaction mixture monotonously increases to steady state level, since $dG/dt \geq 0$. From (2.50) and (2.51) we obtain the derivative of the *net* entropy of an open system, $\sigma(t)$,

$$\sigma(t) = \frac{dS}{dt} = -J_0 R\,e^{-k_0 t}\ln\left(\frac{S(t)}{P(t)}\right). \qquad (2.52)$$

$\sigma(t)$ is the sum of two terms: the *internal* entropy production, $\sigma_i(t)$, occurring in the course of the chemical transformation $S \Rightarrow P$; and *entropy flow*, $\sigma_e(t)$, due to material and heat exchange with the environment: $\sigma(t) = \sigma_i(t) + \sigma_e(t)$. The thermodynamics of irreversible processes enables us to calculate the steady state entropy production, $\sigma_i(\infty) = T^{-1}\mathscr{A}J \geq 0$, where $\mathscr{A}$ is the chemical affinity and J is the rate of chemical reaction. For the reaction mixture of P and S, the chemical affinity equals

$$\mathscr{A}(t) = \mu_S(t) - \mu_p(t) = \mathscr{A}_0 + RT\ln(1 + \alpha) - RT\ln(1 + \beta),$$

where $\mathscr{A}_0 = \mathscr{A}(0)$. In steady state we obtain

$$J_0 = k_0[\phi - 1]\frac{P_0 S_0}{P_0 + \phi S_0}, \qquad (2.53)$$

where $\phi = \exp[(\mathscr{A} - \mathscr{A}_0)/RT]$. Using linear approximation corresponding to the condition $(\mathscr{A} - \mathscr{A}_0)/RT \ll 1$, we finally obtain the relationship between the flux J and the chemical force $\mathscr{A}$

$$J = k_0 \frac{S_0 P_0}{RT(S_0 + P_0)}[\mathscr{A} - \mathscr{A}_0] + j_0, \qquad (2.54)$$

where the parameters $\mathscr{A}_0 = RT \ln[k_+ S_0/k_- P_0]$ and $j_0 = k_+ S_0 - k_- P_0$ are determined by the initial conditions at the moment of switching the flux J. In the general case, $j_0 \neq 0$ and, thus, $\mathscr{A}_0 \neq 0$. Linear approximation gives $\mathscr{A}_0 \cong RT(j_0/k_- P_0)$. Finally, we obtain

$$J = \frac{k_0 S_0 P_0}{RT(S_0 + P_0)}\mathscr{A} - \frac{j_0^2}{k_-(S_0 + P_0)}. \qquad (2.55)$$

If the initial state is an equilibrium one ($j_0 = 0$), $J = [k_0 S_0 P_0/RT(S_0 + P_0)]\mathscr{A}$. In the general case ($j_0 \neq 0$), there is the deviation from simple proportionality between the J and $\mathscr{A}$ values. Since $J > 0$, steady state concentrations of the reagents correspond to the positive chemical affinity $\mathscr{A} = [RT(S_0 + P_0)/k_0 S_0 P_0]J + \mathscr{A}_{th} \geq \mathscr{A}_{th}$ (see (2.55)). Here, $\mathscr{A}_{th} = (\mathscr{A}_0 j_0)/(k_0 S_0)$ is the threshold value of chemical affinity. If in the initial state the system is shifted from the equilibrium to the left side ($\mathscr{A}_0 > 0$), then $\mathscr{A}_0 > \mathscr{A}_{th}$, and, thus, it is not necessary to overcome the threshold $\mathscr{A}_{th}$ in order to support the steady state flux J. If $\mathscr{A}_0 < 0$, the flux J through the system leads to increasing steady state affinity up to the value $\mathscr{A} \geq \mathscr{A}_{th}$.

After the abrupt and simultaneous ceasing at $t = t_0$ of the input and output fluxes ($J = 0$), the system will exponentially relax to the equilibrium concentrations of the reacting particles, S_{eq} and P_{eq},

$$S(t) = S_{eq} + [S_0 + (1 - e^{-k_0 t_0})J_0/k_0 - S_{eq}]\exp[-k_0(t_0 - t)], \quad (2.56)$$

$$P(t) = P_{eq} + [P_0 - (1 - e^{-k_0 t_0})J_0/k_0 - P_{eq}]\exp[-k_0(t_0 - t)]. \quad (2.57)$$

In the course of relaxation to the equilibrium there occurs a stoichiometric transformation of S molecules into P molecules. We see that the conversion factor, characterizing the formation of B molecules, depends on the prehistory of creating the nonequilibrium state. The stoichiometric ratio could be very low ($\eta = K/(1 + K) \ll 1$, when $K \ll 1$) if the system was shifted from the equilibrium by the addition of the substrate, S, to the equilibrium reaction mixture. Meanwhile, for the same value of the equilibrium constant K, *the stoichiometry of transformation becomes high ($\eta = 1$) if the initial state of the system was created by the simultaneous addition of S and withdrawal of P.*

We want to emphasize the latter point in order to avoid any misunderstanding that might occur with the use of formal thermodynamic relationships in open systems. Let us determine the chemical affinity as $\mathscr{A}^* = -(\delta G/\delta \xi)_{p,T}$, where δG is the change in free energy corresponding to the $\delta \xi$ completed acts of chemical reaction. In an open system, however, such formal determination of the affinity can lead to misunderstanding. If J is a steady flux through the system, $\delta\xi = J\delta t$. We might, therefore, formally determine the chemical affinity as $\mathscr{A}^* = -J^{-1}(\delta G/\delta t)_{p,T}$. In the steady state $(\delta G/\delta t)_{p,T} = 0$, and $\mathscr{A}^* = 0$. On the other hand, for any positive value of the flux J the calculated value of chemical affinity $\mathscr{A}$ in a steady state is always positive, $\mathscr{A} > \mathscr{A}_{th} \geq 0$, while the steady state value of $\mathscr{A}^*$ is always zero. This apparent controversy can be easily explained. In an open system the affinity value determined as $\mathscr{A}^* = -J^{-1}(\delta G/\delta t)$ is not a proper measure of thermodynamic force creating the flux through the reaction mixture. As a matter of fact, the "driving force" for performing the energy-unprofitable transformation $S \rightarrow P$ arises as a result of the material exchange with the external medium—steady state inputting S and outputting P from the reaction vessel. The value of chemical affinity $\mathscr{A}$ is an adequate measure of the driving force, that ensures the spontaneous chemical transformation $S \rightarrow P$, but only in the absence of the external force, that takes place after stopping material exchange with the environment. As we have mentioned above, the $\mathscr{A}$ value determines "internal" entropy production. In the linear approximation

$$\sigma_i = T^{-1}\mathscr{A}J \geq 0.$$

Let us further consider the relationships between the chemical affinities and the flux J through two consecutive reactions

$$J(A) \Rightarrow A \underset{k_{-1}}{\overset{k_1}{\rightleftarrows}} B \underset{k_{-2}}{\overset{k_2}{\rightleftarrows}} C \Rightarrow J(C). \tag{2.58}$$

Let A_0, B_0, and C_0 be the initial concentrations of the components, and let $J(A) = J(C)$. Deterministic behavior of this system can be described by the following set of differential equations:

$$dA/dt = J - k_1 A + k_{-1} B,$$

$$dB/dt = k_1 A - k_{-1} B - k_2 B + k_{-2} C, \tag{2.59}$$

$$dC/dt = k_2 B - k_{-2} C - J.$$

If the initial state of a system is in equilibrium, the steady state concentrations, that are established after switching on the input and output fluxes, can be expressed as follows:

$$A = A_0(1 + \alpha), \qquad B = B_0(1 + \beta) \qquad \text{and} \qquad C = C_0(1 + \gamma). \tag{2.60}$$

Here

$$\alpha = \frac{J}{R_0}\lambda_1, \qquad \beta = \frac{J}{R_0}(-K_1^{-1}\lambda_1 + K_2\lambda_2) \qquad \text{and} \qquad \gamma = -\frac{J}{R_0}\lambda_2,$$

where $K_1 = k_1/k_{-2}$ and $K_2 = k_2/k_{-2}$ are equilibrium constants for the partial reactions and

$$\lambda_1 = \frac{k_2 + k_{-2} + k_{-1}}{k_{-1}k_{-2}}, \qquad \lambda_2 = \frac{k_1 + k_{-1} + k_2}{k_1 k_2}.$$

Under steady state conditions that are not too far from equilibrium (i.e., in Onsager's case of linear approximation), the chemical affinities of partial reactions $\mathscr{A}_{AB} = \mu_A - \mu_B$ (reaction $A \Leftrightarrow B$) and $\mathscr{A}_{BC} = \mu_B - \mu_C$ (reaction $B \Leftrightarrow C$), as well as the affinity of resulting reaction $A \Leftrightarrow C$, $\mathscr{A}_{AC} = \mu_A + \mu_C = \mathscr{A}_{AB} + \mathscr{A}_{BC}$, can be easily expressed (in RT units):

$$\mathscr{A}_{AB} = \kappa \frac{J}{R_0}, \qquad \mathscr{A}_{BC} = \kappa \frac{J}{R_0} \cdot \frac{f}{K_1 K_2}, \qquad \mathscr{A}_{AC} = \kappa \frac{J}{R_0} \left(1 + \frac{f}{K_1 K_2}\right), \quad (2.61)$$

where $R_0 = A_0 + B_0 + C_0$, $\kappa = (1 + K_1 + K_1 K_2)/k_1$ and $f = k_1/k_{-2}$.

The steady state flux J through a series of consecutive reactions having a common intermediate thus appears to be proportional to the chemical affinities of the reaction mixture for the partial ($\mathscr{A}_{AB}$ and $\mathscr{A}_{BC}$) and overall ($\mathscr{A}_{AC}$) chemical reactions

$$J = L_{AB}\mathscr{A}_{AB} = L_{BC}\mathscr{A}_{BC} = L_{AC}\mathscr{A}_{AC}. \quad (2.62)$$

This result is in the frame of the traditional approach of phenomenological linear thermodynamics of irreversible processes. It follows from (2.61) that in the case of two consecutive reactions the ratio of the affinities for two partial reactions, $\mathscr{A}_{AB}/\mathscr{A}_{BC} = f/K_1 K_2$, is determined not only by thermodynamic properties of the system (characterized by the equilibrium constants K_1 and K_2), but also depends on the relationship between the rate constants of forward and backward processes (parameter $f = k_1/k_{-2}$), i.e., depends on the kinetic properties of the chemical system. This result may be of certain interest for the interpretation of experimental data concerning the energy transduction in certain biochemical processes.

In this section we have considered the two simplest examples using the deterministic description of chemical reactions. This approach is adequate but only in the so-called thermodynamic limit when we can neglect the discrete nature of the processes considered, as well as the fluctuations of the reactants. Rigorous consideration of these processes becomes possible within a stochastic approach to the description of chemical reactions (for references, see the excellent review by McQuarrie [20]). For the sequence of monomolecular reactions in open systems with an arbitrary number of intermediates, the problem has been investigated in depth by Nicolis and Babloyantz [31], Ishida [32] and other authors (see, for references, [33]). The stochastic approach, however, faces serious analytical difficulties for more complex systems (for instance, the bimolecular reaction $A + B \Leftrightarrow C$). Some unusual properties of this reaction in small volumes, associated with enormously large fluctuations, will be considered in Chapter 3.

References

1. E. Schrödinger (1945), *What Is Life? The Physical Aspects of the Living Cell*, Cambridge University Press, Cambridge.

2. L.A. Blumenfeld (1989), *Life and Science (USSR)*, No. 10.

3. L.A. Blumenfeld (1981), *Problems of Biological Physics*, Springer-Verlag, Heidelberg.

4. Th. De Donder (1927), *L'affinite*, Gauthier-Villars, Paris.

5. F.H. Johnson, H. Eyring, and M.J. Polissar (1954), *The Kinetic Basis of Molecular Biology*, Wiley, New York.

6. P.W. Atkins (1983), *Physical Chemistry*, Oxford University Press, Oxford.

7. N.G. Van Kempen, *Stochastic Processes in Physics and Chemistry*, North-Holland, Amsterdam.

8. R. Welch (1985), *J. Theor. Biol.* **114**, 433–446.

9. J.W. Gibbs (1902), *Elementary Pinciples in Statistical Mechanics*, Yale University Press, New Haven.

10. S. Glasston, K.J. Laidler, and H. Eyring (1940), *Theory of Rate Processes: The Kinetics of Chemical Reactions, Viscosity, Diffusion, and Electrochemical Phenomena*, McGraw-Hill, New York.

11. K.J. Laidler (1965), *Chemical Kinetics*, 2nd edition, McGraw-Hill, New York.

12. H. Eyring, J. Walter, and G.E. Kimball (1945), *Quantum Chemistry*, Wiley, New York.

13. B.E.C. Banks and C. A. Vernon (1970), *J. Theor. Biol.* **29**, 301–326.

14. Ya.S. Lebedev, Yu.D. Tsvetkov, and V.V. Voevodskii (1960), *Kinet. Katal. (USSR)* **1**, 496.

15. L. Onsager (1931), *Phys. Rev.* **37**, 405–426.

16. L. Onsager (1931), *Phys. Rev.* **38**, 2265–2279.

17. I. Prigogine (1967), *Introduction to Thermodynamics of Irreversible Processes*, Wiley, New York.

18. R. Balescu (1975), *Equilibrium and Non-equilibrium Statistical Mechanics*, Wiley–Interscience, New York.

19. A.N. Shilov (1905), *On the Coupled Oxidation Reactions*, Moscow.

20. D.A. McQuarrie (1967), *Appl. Probab.* **8**, 1–66.

21. C.W.F. McClare (1971), *J. Theor. Biol.* **30**, 1–34.

22. T.L. Hill (1977), *Free Energy Transduction in Biology*, Academic Press, New York.

23. A. Kachalsky and P.F. Curran (1965), *Non-equilibrium Thermodynamics in Biophysics*, Harvard University Press, Cambridge, MA.

24. H.V. Westerhoff and K. van Dam (1979), In: *Current Topics in Bioenergetics* (D. Rao Sanadi, ed.), Vol. 9, Academic Press., New York. pp. 1–62.

25. H.V. Westerhoff and K. van Dam (1987), *Thermodynamics and Control of Biological Energy Transduction*, Elsevier/North-Holland, Amsterdam.

26. S.R. Caplan and A. Essig (1983) *Bioenergetics and Linear Nonequilibrium Thermodynamics*, Harvard University Press, Cambridge, MA.

27. P. Glansdorf and I. Pigogine (1971), *Thermodynamic Theory of Structure, Stability and Fluctuations*, Wiley–Interscience, London.

28. H. Rottenberg (1979), *Biochem. Biophys. Acta* **549**, 225–283.

29. J.W. Stucki (1980), *European J. Biochem.* **109**, 269–283.

30. T.L. Hill and E. Eisenberg (1981), *Quart. Rev. Biophys.* **14**, 463–551.

31. G. Nicolis and A. Babloyantz (1969), *J. Chem. Phys.* **51**, 2632–2637.

32. K. Ishida (1973), *J. Theor. Biol.* **40**, 301–327.

33. A.T. Bharucha-Reid (1960), *Elements of the Theory of Markov Processes and Their Applications*, McGraw-Hill, New York.

CHAPTER 3

Molecular Machines: Mechanics and/or Statistics?

What Are Molecular Machines and Whether or not a Classical Thermodynamic Approach Is Valid for the Description of Biological Systems

> The problem we have to consider is the molecular nature of living things, and, particularly, whether such systems can work the same way as ordinary chemical machines.... Only by re-examining some of our fundamental beliefs will it be possible to resolve the problems which exist at present in bioenergetics.
>
> (C.W.F. McClare, Chemical machines, Maxwell's demon and living organisms (1971), *J. Theor. Biol.* **2**, 1–34.)

3.1. The Second Law of Thermodynamics and Its Application to Biochemical Systems

To understand the main features of the mechanisms of energy transduction at the molecular level we need, first of all, to get an answer to the question: How the laws of thermodynamics, including the Second Law, can be applied to explain performing work by one macromolecule acting individually and independently on the states of other ones? The interest in this subject had been suggested to one of us (L.A.B.) as early as 1971–1972, by the analysis of certain biophysical aspects of enzymes functioning [1, 2]. The scrutiny of the "mechanical" properties of macromolecules had led to the formulation of the new concept of enzyme catalysis, called the relaxation concept. At the same time, McClare independently began to analyze the operation of macromolecules considering them as molecular machines [3–6]. We want to start our analysis of this problem by reviewing in brief some principle notions in theoretical bioenergetics that had been put forward by McClare, although, it would be much more interesting and useful for the reader to read his original works [3–6]. Many of McClare's original and clear ideas were ahead of conventional concepts in the realm of biothermodynamics, and were probably not appreciated in full by the majority of biophysicists and biochemists. The main points and conclusions of this chapter are consonant with McClare's ideas.

McClare in his pioneering work had scrutinized the problem as to whether or not conventional chemical machines can be used in living organisms [3–6].

He analyzed the applicability of thermodynamics for describing biological systems. Of course, there is no doubt that the First and Second Laws of Thermodynamics hold true, and there are no reasons to refute them. However, considering the peculiarities of biological systems, McClare extended classical thermodynamics statements to the molecular level, pointing out that living systems pose unique thermodynamic problems. In his approach to the solution of these problems, McClare introduced time, τ, into the fundamental statement of the Second Law of Thermodynamics. This time scale, being determined by the intrinsic properties of the molecules, is a characteristic property of the system. Reformulating the classical statement of the Second Law, McClare indicated how it could be valid at the molecular level.

Kelvin's classical statement of the Second Law—"*It is impossible to devise any engine which, working in a cycle, shall produce no effect other than the extraction of heat from the reservoir and the performance of an equal amount of mechanical work*" (cited in [7]). According to McClare [3], this formulation is inadequate in biology. The classical statement of the Second Law does not contain time. Having been stated more than 100 years ago, this formulation did not, of course, take into account the internal peculiarities of molecules nor their atomic constitution. Meanwhile, the atomic structure of biopolymers cannot be ignored in explaining the energetics of molecular machines. McClare's statement of the Second Law introduces the characteristic time, τ. In this way, in a rather common form, the molecular properties of the system could be taken into account. Introducing a clear definition of heat and work at the molecular level, McClare modified the classical statement of the Second Law to biological systems. In order to avoid semantic difficulties, he used a new term "stored energy," thus subdividing the energies contained in a system into "*stored energy*" and "*thermal energy*." Let the system be at equilibrium at temperature T_1, and τ is the characteristic time interval of the processes considered. *Thermal energies* were defined as those "*which exchange with each other and reach equilibrium in the time less than τ so that they obey a Boltzmann distribution characterized by T_1.*" Stored energies were defined as those "*which remain in a different distribution for a time longer than τ; either in a distribution characterized by a higher temperature T_2, or such that higher energy states are more populated than states of lower energy.*" Also, *stored energy* means "*any form of energy which does not exchange with translational, rotational, and vibrational energies which normally constitute heat.*"

By analogy with Kelvin's statement, McClare reformulated the Second Law: "*It is impossible to devise an engine, of any size whatever, which, acting in a cycle which takes time τ, shall produce no effect other than extraction of energies, which have equilibrated with each other in a time less than τ, from a reservoir at one temperature and the conversion of these energies into a form in which they would be stored for longer than τ; either at a higher temperature, or in a population-inversion.*"

A short statement of the Second Law was formulated as follows: "*Useful work is only done by a system when one form of stored energy is converted into another."*

In this context, performing useful work by the molecular machines of living cells implies that the conversion of one form of stored energy into another does not occur via thermal exchange, as takes place in all conventional macroscopic chemical machines. This statement not only defines useful work at the level of individual molecules, but also points out that the difference between work and heat disappears at the molecular level [3–5].

We will now consider some other problems that are relevant to the processes of energy transduction in biochemical systems:

(i) the definition of thermodynamic functions that are pertinent to characterizing the energetic status of different transient states of individual macromolecular enzyme complexes;

(ii) the localization of the crucial steps in the enzyme turnover cycle that might be responsible for energy transduction; and

(iii) What term of free energy changes, enthalpic or entropic, can be used for performing work at the molecular level?

Discussing the problem of the adequate use of the Second Law for biological systems, McClare indicated that from the point of conventional equilibrium thermodynamics, entropy is a *macroscopic* function of the system's state. Entropy change determines the direction of spontaneous irreversible processes in the whole system. At the "intermediate" level of structural organization, described within the approach of statistical mechanics, entropy is a *mesoscopic* value which is determined by the probability partition function. On the other hand, because of the reversibility of physical processes at the microscopic level, entropy cannot be a *microscopic* value. This statement had been clearly argued as early as 1912 by Paul and Tatyana Ehrenfest [8]. Following this line of argument, McClare deduced that entropy cannot be a characteristic function of molecules at the microscopic level.

In accordance with the widely accepted traditional point of view, concluded McClare, it might mean that the Second Law has a statistical nature, only being applied to the systems consisting of a large number of particles. The simplest example of this statistical ensemble is a gas confined to a macroscopic container. In the thermodynamic limit, corresponding to a large number of particles in a system, the fluctuations of macroscopic values are small. In the frame of statistical mechanics formalism, thermodynamic potentials, characterizing macroscopic properties of a system in the equilibrium state, can be expressed through a calculation statistical sum [9]. Formally, the statistical sum can be determined regardless of the number of particles in the system. For systems consisting of a small number of particles, however, the fluctuations of macroscopic parameters (such as concentrations of particles, and thermodynamic potentials) can be too high. This may be the case for some chemical mixtures confined in small volumes (an example of this kind, which is relevant to the energetics of chemical reactions occurring within small vesicles of energy-transducing organelles, will be considered below, Section 3.3).

Bearing in mind the reversibility of the processes at molecular level, and following McClare's arguments [3], it might seem at first glance that there could arise an apparent violation of the Second Law if the system contains single, or a few, molecules. We want to emphasize, however, that this statement may be reasonable for only relatively small molecules, but is not a common case for macromolecular systems. We have to take into account that each individual macromolecule itself can be considered as a statistical system [10]. Moreover, it is individual molecules, or complexes consisting of a relatively small number of macromolecules, that usually play the role of functional units in a living cell. Some organelles, cells, and even organisms, are often so small that each of them contains only a few molecules of enzymes and coenzymes. Thus, for an adequate thermodynamic description of biochemical processes in complex biological structures, there is no reason to consider the behavior of a large number of spatially separated enzyme molecules, organelles, or cells in order to have a statistical ensemble of particles.

As for macroscopic systems, the transition from one state of an individual biopolymer molecule (or of its complex with low-molecular ligands) to another state can be characterized by the change in thermodynamic potentials, such as free energy, and this transition would therefore correspond to a certain change in entropy. The behavior of macromolecules, including their structural changes, may have an irreversible character (for instance, heat denaturation of proteins and nucleic acids). The direction of those processes in biopolymers must be determined by the changes in their thermodynamic parameters. In other words, the behavior even of a single macromolecule can be described by the laws of statistical thermodynamics.

Let us now consider in more detail the definitions of free energy related to enzyme reactions. Under steady state conditions, functioning enzyme complexes undergo cyclic transitions between a number of different states. These states can differ in the composition of a complex (enzyme molecule with ligands, substrates, products, low-molecular affectors, etc.), as well as in the conformations of an enzyme molecule. The complex's transitions are coupled with the chemical transformations of substrate molecules, the processes of association–dissociation of substrates and products, active transport of various substances, muscle contraction, etc. Most of these processes are of course associated with the energy transduction from one form to another.

There are two extreme approaches for the description of the energy transduction of these processes:

(i) theoretical consideration of the processes on the molecular level based on the knowledge of atomic details of concrete molecular structure [11–17]; and

(ii) a completely phenomenological description based on the formalism of the Onsager's–Prigogine nonequilibrium thermodynamics, including its generalization for the description of nonlinear systems that are very far from equilibrium [18–30].

The former approach, i.e., using an *ad hoc* principle, can be realized only in a limited number of cases because of insuperable technical problems (it needs the knowledge of a tremendous number of atomic details of macromolecules and the mechanisms of catalyzed reactions). One of the most interesting results among theoretical investigations of this kind was the molecular dynamics simulation of proton transport in the active center of a lysozyme, performed by Warshel [17]. The pure phenomenological approach appears to be very useful for analyzing thermodynamic relations in bioenergetics (for references, see [25–30]). However, this approach cannot clarify detailed mechanisms of chemical transformations. In many cases concerning energy transduction in macromolecular and multienzyme complexes, the most productive appears to be the intermediate approach. This approach involves kinetic analysis with given rate constants attributed to the transformations between discrete macromolecular states of the enzyme–ligand complexes. The powerful analytical tool for such kinds of theoretical description have been introduced by King and Altman [31]. This so-called diagram method was rediscovered and significantly extended by Hill who widely used it for the description of thermodynamic cycles and fluxes, transport processes, and coupling chemical processes (for comprehensive references, see [26, 27]).

In the ensemble of equivalent particles (dilute solution of macromolecules or their complexes) each individual macromolecule unit is practically independent of the others. We can also say that a single macromolecular complex itself represents a statistical system characterized by a number of discrete states. For adequate characterization of energy states of *individual macromolecules and their complexes* with substrates, products, and ligands, Hill introduced the term "*basic free energy.*" This term relates to levels of free energy of the individual states of macromolecular complexes at fixed concentrations of substrates, products, and ligands in the surrounding medium. This characteristic function differs, of course, from the overall Gibbs free energy of the whole system ("*gross free energy*"), related to the ensemble of macromolecules and their low-molecular ligand (for details, see [26, 27, 32]).

To emphasize why it is important to deal with the levels of basic free energy when we speak of the energetics of enzyme processes let us consider, for instance, the population of energy-transducing organelles (mitochondria, chloroplasts, or chromatophores). The ability of certain ATPase molecules to catalyze the ATP hydrolysis does not immediately depend on the states of similar enzyme molecules and low-molecular components (substrates, products, and ligands) located in other compartments or other cells. This situation may be of special importance in the case of heterogeneous populations of vesicles within one and the same organelle (for instance, thylakoids of grana and stroma in chloroplasts, see Chapter 5 for more details).

The behavior of the total system will also be dependent on the interactions within a large statistical ensemble of such macromolecular units. In any real system even spatially separated macromolecules, say ATPase complexes, may have interconnections via common metabolites. For this reason, the free energy of the total system will include not only the basic free energy terms related to macromolecular complexes, but also the concentration-dependent terms related to free (unbound) low-molecular metabolites. The overall Gibbs free energy of the whole system, designated as "*gross free energy*" [26, 32], can be calculated from the reagent's partition function related to the *whole* ensemble of macromolecules and their low-molecular ligand. According to [26, 32], two terms of chemical potential differences, corresponding to "gross" and "basic" free energy changes and relating to the transition from one state to another, $\Delta\mu_{ij}$ and ΔA_{ij}, are connected through a logarithmic term of purely entropic origin

$$\Delta\mu_{ij} = \Delta A_{ij} + k_B T \ln(p_i/p_j),$$

where p_i and p_j are the probabilities of the enzyme states "i" and "j."

What is the driving force for the enzyme catalyzed coupling energy-donating and energy-accepting processes? From the conventional thermodynamic point of view, this coupling is governed by the mass action law. The chemical affinities of the substrate and product molecules, related to the volume of a whole reaction mixture, will determine the overall direction of the enzyme catalyzed processes. In the frame of this approach, the free energy derived from the energy-donating reaction, say hydrolysis of ATP, should be obviously related to the overall change in the free energy of the whole system ("*gross*" free energy change). After the complete turnover, the enzyme molecule returns to the initial state, its energy level does not change, and thus all terms related to the enzyme *basic* free energy changes will cancel. This point was clearly expressed by Hill: "net reaction (positive flux) always occurs in a *downhill* direction with reference to a set of gross free energy levels. This is _not_ true of the invariant basic free energy levels" [26].

In order to obtain an answer to the questions about the role of the enzymes in energy transfer, we need to refer to the elementary processes occurring at the level of the individual macromolecules (or their complexes). More than two decades ago the problem of energy storage in a single molecule (for instance, the ATP molecule), and its utilization by a single enzyme molecule, regardless of the functioning of the other components of the ensemble, was actively discussed in the biochemical literature (for details, see [33–36], and other references in [2]). In this connection the relevant question arises: Is it possible to localize the crucial steps in the enzyme cycle that are responsible for the free energy transfer? As a matter of fact, this question concerns the dilemma: What properties, statistical or mechanical, determine the behavior of macromolecular complexes?

A purely statistical point of view on this subject has been explicitly expressed by Hill [26, 37] who analyzed free energy transduction in biochemical cycles. Later this problem was scrutinized and elaborated on at length in [27]. As Hill argued, "free energy transduction in cyclical biochemical systems should be attributed, in general, to complete cycles and not to individual steps or transients belonging to the cycles," therefore, it is a misunderstanding "to select the one particular enzymatic step or transition at which the free energy transduction supposedly occurs" [37]. As an illustrative model he referred to the hypothetical process of the net active transmembrane transport of a ligand L from outside to inside ($L_o \rightarrow L_i$), against its concentration gradient. The source of energy for this unprofitable "uphill" transport was assumed to be the free energy of ATP hydrolysis catalyzed by the membrane enzyme complex, E. The above-cited conclusions were founded on the analysis of a kinetic diagram and a hypothetical set of enzymatic "basic" free energy levels corresponding to the net reaction, ATP + $L_o \rightarrow$ ADP + P_i + L_i. The levels of "basic" free energy were assumed to be determined by the apparent first-order (or pseudo first-order) forward and backward rate constants (α_{ij} and α_{ji}) in the cycle, and thus corresponding to a set of discrete kinetic states. With the transition from the enzyme state "i" to "j," the basic free energy change can be defined as $\Delta A_{ij} = k_B T \ln(\alpha_{ij}/\alpha_{ji})$. After completion of the cycle and returning the enzyme to its original state, the concentrations of the substrates and products of the energy-donating reaction (ATP $\rightarrow$ ADP + P_i), as well as that of energy-accepting reaction ($L_o \rightarrow L_i$), will change. Therefore, there will be a change in the level of the system's "gross" free energy, while net "basic" free energy changes will drop out for the complete cycle. Hill had come to the same conclusions in his analysis of the energetics of the cross-bridge cycle and sliding of filaments in muscles (Chapter 5 in [26]).

We think that the very conclusion, that it is impossible to localize the crucial stages (or steps) of the enzymatic cycle responsible for energy transduction, was initially (*a priori*) hidden in the purely thermodynamic (statistical) approach to the problem. As we have mentioned above, in many aspects the behavior of macromolecules resembles that of mechanical systems. Meanwhile, the conventional thermodynamic approach does not take into account the time scales of the processes considered. Characteristic time scales of the dynamic processes, such as say the conformational relaxation initiated by the substrates or other ligands binding and/or releasing, are determined by the intrinsic properties of the concrete enzyme molecules. These time scales, however, are lacking in the thermodynamic treatment of equilibrium or steady state processes. As we have mentioned above, in rather general form, this notion has been clearly formulated by McClare [3–6] and Blumenfeld [1, 2]. At any rate, a purely thermodynamic (statistical) approach implies the establishment of at least the thermal equilibrium for every macromolecular complex. Only in this case can we speak of free energy in its conventional sense. Otherwise, the existence of certain rather long-living states of macromolecules with the "excited selected degrees of freedom" implies the lack of thermal equilibrium with the surrounding medium. In these nonequilibrium states, for a certain time, τ, at least part of the energy is conserved in its "noble" form (the so-called "stored energy"); these excited degrees of freedom do not reach equilibrium with the thermal degrees of freedom within this interval of time. In this case, for an adequate description of the energy status of the macromolecules we must use the *full energy term* but *not* *free energy*.

3.2. Energy-Transducing Molecular Machines

Starting from the extended statement of the Second Law (that includes the characteristic time of molecular processes), McClare [3–6] deduced that there are two general ways of obtaining useful work from the chemical reactions:

(i) a "*constrained equilibrium*" mechanism; and
(ii) a "*molecular energy*" mechanism.

Independently of McClare, similar ideas have been put forward by one of us [1, 2] who elaborated on the concept of molecular machines for enzyme catalysis and energy transduction in chemical and biochemical systems. A central point of this concept was founded on the notion of the crucial role of the conformational relaxation of biopolymers in catalyzing biochemical processes. We will consider the application of this relaxation concept to enzymology and bioenergetics in Chapters 4 and 5. We now focus our attention on the general principles of energy transduction by *molecular machines*.

3.2.1. Macroscopic Machines

The "*constrained equilibrium*" mechanism of energy transduction is usually employed by conventional macroscopic machines such as heat engines, which are able to transform energy released in the course of a chemical reaction (fuel combustion) into mechanical work. The very possibility of the efficient functioning of these machines is determined by the constructional peculiarities of energy-transducing devices: any conventional chemical machine represents a special *construction* with a limited number of mechanical degrees of freedom. This construction *constrains* the behavior of the statistical part of a device consisting of a large number of various particles (e.g., gas molecules confined to the cylinder of a steam engine, reacting compounds of fuel inside the cylinder of an internal combustion engine, the ions in the compartment of a galvanic cell, an electron "gas" in metal wires connecting an electric generator with a load, etc.).

Isothermal expansion of an ideal gas inside a cylinder with a traveling piston (Fig. 3.1) represents a typical step of the simplest macroscopic machine. Gas molecules are confined to the constraining part of the machine (cylinder), and the functioning of this device is associated with the movement of its mobile part (piston) corresponding to its traverse along one selected

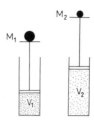

Fig. 3.1. Sketch of the device for performing work using isothermal expansion of the ideal gas.

degree of freedom. This machine can perform mechanical work: the thermal energy of a gas is transformed into the potential energy of a lifted load due to the "excitation" of the mechanical degree of freedom. The maximal efficiency of energy transduction can be attained during a reversible (quasi-reversible) mode of functioning, corresponding to the infinitesimally slow movement of a piston.

Analyzing the problem of doing work by so-called *"entropic"* machines, one of us has pointed out that there are only two kinds of such machines related to the "constrained equilibrium" mechanism of energy transduction. Speaking of entropic machines, we mean such devices that can use an entropic form of free energy change in order to perform useful work. Similar to the above-considered case of the entropic mechanism of coupling between energy-donating and energy-accepting reactions (Section 2.4), entropic machines extract heat from the thermostat. "Entropic" machines of the first kind are able to transform thermal energy into useful work by *exciting the mechanical movement of macroscopic parts of the machine due to kinetic energy of molecules.* "Entropic" machines of the second kind use the constructional peculiarities of the device that *facilitate the selection of "hot" particles which are able to overcome a potential barrier* lying in the path of chemical transformation. The mechanism of the first kind can be illustrated by the model of isothermal expansion of an ideal gas confined to the cylinder with a frictionless moving piston. The mechanism of the second kind is illustrated by in the case of an electrolyzer and concentrational galvanic cell [38].

Let us consider one step of the functional cycle of a mechanical entropic machine—expansion of an ideal gas confined to a cylinder with a movable piston (Fig. 3.1). The external load M applied to the top of the piston is assumed to be initially equilibrated by the gas pressure, $p = nRT/V$, where V is the initial volume of gas in the cylinder, T is the gas temperature in degrees Kelvin, R is the universal gas constant, and n is the number of moles of the gas. The system is assumed to be in contact with a thermostat. On gradually reducing the weight of the external load from M_1 to M_2, gas will expand, lifting the load and thus performing mechanical work, ΔW. If the motion of a piston occurs slowly enough, the temperature of the gas will remain constant due to heat exchange with the thermostat. In the course of this slow, and thus *reversible*, expansion, gas will perform work $\Delta W = nRT \ln(V_2/V_1)$, where V_1 and V_2 are the initial and final volumes occupied by gas in a cylinder. This work is equal to the decrease in gas free energy, $\Delta G = -T\Delta S = -nRT \ln(V_2/V_1)$, determined solely by the entropic term of free energy. There are no other processes but heat flow that could cause the change in enthalpy of an ideal gas; the isothermal expansion of gas would not lead to its enthalpy change. The cooling of expanding gas (that would take place due to losing kinetic energy by gas molecules colliding with the wall of a traveling piston for an isolated system) is prevented by heat flow from a thermostat, ΔQ. Mechanical work, ΔW, at this stage will thus be performed due to heat energy consumed by the gas. Meanwhile, the ideal gas in the cylinder, used as a "working body" of the machine, will not change its enthalpy in the course of isothermal expansion. The efficiency of performing mechanical work (if friction can be neglected) at the stage of the slow (reversible) expansion of the gas appears to be maximal, $\eta = \Delta W/\Delta Q = 1$.

At this step of machine functioning, the transformation of thermal energy acquired from the thermostat into the potential energy of the lifted weight becomes possible due to one important property of the mechanical device. The machine containing a large number of particles (gas inside a cylinder) represents a construction having only one mechanical degree of freedom determined by the position of the piston. On reducing the load M, the piston is able to move along the cylinder using the kinetic energy of gas molecules bombarding the piston wall. Each individual molecule of gas elastically colliding with the wall loses relatively small amounts of kinetic energy, but being gathered from a large ensemble of gas molecules this energy appears to be finally transformed into potential energy of the lifted weight. Due to constructional peculiarities of the system this form of energy can be stored for a long time.

Performing work by conventional macroscopic machines (e.g., a heat engine), acting according to a "constrained equilibrium" mechanism, is most efficient if the machine operates slowly enough, thus providing the conditions for the establishment of thermal equilibrium at the crucial working stage of the machine cycle. Speaking of "constrained equilibrium machines" we imply that the working stroke stage of these devices includes heat exchange with the thermostat. A typical example of such a process is the Carnot cycle. Rapid operation of such macroscopic machines would limit their efficiency. Being essentially equilibrium in character, the "constrained equilibrium" mechanism cannot be used efficiently for energy transduction in biological processes. Physical reasons which preclude using conventional chemical machines functioning under quasi-equilibrium conditions in biological systems are discussed below.

3.2.2. What Are Molecular Machines? Reversibility of Energy-Transducing Devices and the Problem of the Optimal Functioning of Molecular Machines

Doing useful work, or performing energy transduction in biological systems, was identified by McClare [3] with the "*process of converting one form of stored energy into another.*" At the molecular level, this process ought to be put into practice by cyclic operations of special enzyme molecules or their complexes embedded in a membrane, that could be called *molecular machines* [1–6, 38]. The main properties of molecular machines are:

(i) the "individual" working of each machine does not practically depend on the states of other molecular devices of the macroscopic ensemble;
(ii) a fast mode of operation;
(iii) molecular machines are enthalpy-driven constructions;
(iv) entropy production by a molecular machine is a wasteful process; and
(v) conformational flexibility and structural changes of molecular machines play a crucial a role in their functioning.

According to the conventional approach, a thermodynamic process can be performed reversibly if it proceeds at an infinitely slow rate. In contrast to this generally accepted view, McClare stated that at the molecular level, "under appropriate conditions, even very rapid processes can be reversible too" [3]. Having introduced time into the formulation of the Second Law, he was able to indicate the conditions for the reversible operation of molecular machines. This problem is also rigidly associated with the demarcation between thermodynamics and mechanics, or the statistical and mechanical description of macromolecular processes. How do we divide the processes into mechanical and thermodynamic processes? According to McClare ([3], p. 10), "*A thermodynamic process is one in which thermal exchange is a necessary part. A mechanical process is one in which thermal exchange is either irrelevant or at equilibrium.... A thermodynamic process becomes a mechanical process when it is carried out slow enough.*"

In contrast to macroscopic machines, the efficiency of molecular machines would be greater if they operate as rapidly as possible. In this kind of machines, energy stored in a single molecule must be converted into another form (or transferred to another energy-accepting molecule) rather quickly, in order to preclude its dissipation into heat. If stored energy is allowed to reach equilibrium with the environment, then this energy would be wasted. In other words, the act of energy transduction must be completed before thermal exchange would dissipate the stored energy. Also, in contrast to the mechanisms of indirect coupling considered above, the mechanism of direct energy conversion ensures the coupling of energy-accepting and energy-donating processes without equilibrating with the thermal degrees of freedom of the surrounding medium. The characteristic time of heat exchange processes for low-molecular compounds in a homogeneous reaction mixture is about

10^{-12}–10^{-11} s. Very fast relaxation along the thermal degrees of freedom makes improbable the appearance of the appropriate "hot" reacting particles with the excess of energy that occurs as a result of the energy-donating reaction. This is why a very *fast thermal exchange with the surroundings practically excludes the direct transfer of energy* liberated in the course of an energy-donating process to energy-accepting steps of a chemical transformation in a coupled reaction. In the case of macromolecules, this rapid dissipation of energy can be precluded, thanks to the intrinsic properties of macromolecules.

Among the most important functional properties of biopolymers is that many of them demonstrate relatively long-living nonequilibrium conformational states. Characteristic life-times of these states (up to $\tau \cong 10^{-6}$–10^{-3} s, and up to 1 s in several special cases (see Chapters 4 and 5), are much higher as compared with the short time of thermal exchange ($\tau \cong 10^{-12}$–10^{-10} s). We think that this is the reason why energy-transducing macromolecules, operating in nonequilibrium conditions, can perform useful work with high efficiency.

For macroscopic machines, the maximal efficiency of their functioning can be achieved with a reversible (or quasi-reversible) mode of their operation. This implies that for any step of the machine working cycle, the driving force must be as close as possible to the applied external load. In order to perform optimal movement of any part of a mechanical construction (that can be driven, for example, by gas expansion in the cylinder of a heat engine), the force $\vec{F}$ generated by the machine must be as close as possible to the absolute value of the force $\vec{F}_{ext}$ acting from the load; i.e., $\vec{F}(\vec{R}) = -\vec{F}_{ext}$ is the condition for the optimal functioning of the macroscopic machine, where $\vec{R}$ is a generalized coordinate determining the machine configuration. In any real case, $|\vec{F}(\vec{R})| = |\vec{F}_{ext}| + |\varepsilon|$, where $|\varepsilon|$ is the "futile excess" of applying force $\vec{F}(\vec{R})$ over the $\vec{F}_{ext}$. To provide the optimal regime of machine functioning we need $|\varepsilon| \to 0$ in the limit. For macroscopic machines, the conformity of the force created by the machine to the external load can be achieved, for example, by adjusting the appropriate gas pressure acting on the piston of the heat engine, or an appropriate electric current in the electric engine. In order to generate such an optimal, i.e., properly fitted, driving force, every macroscopic machine must have a special device for performing feedback control. This control is achieved on the basis of information about the load (the precise position of the elements of the construction, force value, electric current, etc.).

According to the information theory [39], the information cannot be obtained gratis. Increasing the information is equivalent to decreasing the entropy of the system. This process cannot proceed spontaneously and therefore a "noble" form of energy should be spent to obtain the information. Another line of argument has led Landauer and Bennett to the conclusion that energy should be expended in order to get rid of the excess information in a cyclically operating device [40–43]. In any event, in order to perform the feedback control, even for the frictionless devices, we have *to spend a certain amount of energy to obtain (or to get rid of) the information* about the load position that allows us to adjust optimal force at any stage of the machine operating cycle. Considering the problem of wasting energy for informational reasons, we will discuss below the question of energy payment for information. The scrutiny of energy costs for getting rid of the excess of information does not change the main conclusions.

For macroscopic machines, the energy payment for information can be relatively small as compared with the useful work performed by the machine. However, this is not the case, in general, for machines operating at the level of individual molecules. We will now consider two aspects of the problem of optimal efficiency of molecular machine action:

(i) the loss in the efficiency due to the energy "cost" of information; and
(ii) optimal "fitting" between the force created by the machine and the load at the level of individual molecules.

To clarify these problems, we will analyze below certain highly simplified but quantitative models.

Considering the problem of the reversibility of a biological machines capable of performing mechanical work, Gray [44] concluded that at the molecular level the payment for information is too high to provide a reversible but efficient regime for this kind of machine functioning. For this reason, in general, the reversible mode of molecular machine operation is not optimal.

Following Gray's arguments [44], let us evaluate the energy payment for information necessary for the reversible operation of a machine. It is easy to demonstrate that Gray's main conclusion is correct, regardless of some inaccuracy in the calculation of the conversion factor that has been made in [44].* Again, as an example we consider a mechanical machine—a cylinder with a traveling piston (Fig. 3.1). In order to perform an expansion (or compression) of gas in a cylinder under *quasi-reversible* conditions, the applied external force $\vec{F}_{ext}$ at any position of a piston (coordinate X) must be as close as possible to the force function $F(X)$, i.e., $F_{ext} \cong F(X)$. To obey this condition, and thus to reach optimal operation of the machine, the device requires information about the position of the piston. According to quantum theory, the precision of ascertaining the piston position, X, is limited by Heisenberg's uncertainty relationship $\langle(\Delta p_x)^2\rangle\langle(\Delta X)^2\rangle \geq \hbar^2/4$, where Δp_x is the uncertainty in the impulse of a particle whose position, X, is determined with the uncertainty ΔX; $\hbar = h/2\pi$ is the Planck constant.

If a quantum of light is used for measuring the position of a particle [39], then the energy cost, ΔE, for the determination of X within the limits ΔX is

---

* Calculating the conversion factor θ, Gray has used the formula $\theta = W/(Q + \Delta E)$, where W is the work done by the system, ΔE is the energy cost of information, and Q is the net energy input. However, any device working cyclically has to "pay" the cost of information due to energy inputted, Q. Therefore, the term ΔE should be involved in the Q value and, thus, the conversion factor must be calculated as $\theta = W/Q$.

evaluated by the relationship

$$\Delta E \geq hc/4\Delta X, \tag{3.1}$$

where c is the velocity of light. Increasing the precision of the particle position ($\Delta X \to 0$) increases the energy cost, ΔE. The total energy input into the system, Q, must include the expenditure of energy for obtaining the information, ΔE. The efficiency of useful work done by the system per one cycle of machine operation, η, should tend to its maximal value η_{id} if the machine operates in the reversible mode of action.

As a consequence of the uncertainty in the position of a "piston," the force F_{ext} cannot be exactly equal to the function $F(X)$ at any moment of time, $F(X) = F(X)_{ext} + \Delta F$. We assume that $F(X) \geq F(X)_{ext}$, and that this condition ensures overcoming the external force $F(X)_{ext}$ at any stage of a machine cycle. The deviation of real force developed by the working machine, $F(X)$, from the external force, F_{ext}, can be evaluated as $\Delta F(X) \cong (\partial F(X)/\partial X)\Delta X$. As a result of one cyclic turnover the machine will perform total work W

$$W = \oint F(X)\, dX = W_{ext} + \oint \Delta F(X)\, dX = W_{ext} + \Delta W, \tag{3.2}$$

where $W_{ext} = \oint F(X)_{ext}\, dX$ is useful work. The difference $\Delta W = W - W_{ext}$ is the futile work that represents the "waste" of energy per cycle caused by the nonideal fitting of the forces arising due to the uncertainty in ascertaining the position X. Thus, due to nonideal force–load fitting, the efficiency of machine operation, η, will be lower than the ideal efficiency η_{id}, corresponding to the case of perfect fitting between the force created by a machine, $F(X)$, and an external load, $F(X)_{ext}$. To evaluate $\Delta W = \oint \Delta F(X)\, dX$, we approximate the integral by the sum corresponding to several elementary steps of the machine cycle

$$\Delta W = \sum_i \int_{X_i}^{X_{i+1}} \Delta F(X)\, dX, \tag{3.3}$$

where the symbol "i" relates to the ith step of the machine cycle. Assuming $\Delta X \cong$ const. and $F(X_i) \cong$ const. for any elementary step "i," we obtain

$$\Delta W \cong \Delta X \sum_i |F(X_i)|, \tag{3.4}$$

where ΔX is the uncertainty of a "piston" position.

Let us designate by Q the energy input per one cycle of the machine turnover (e.g., Q equals the energy released in the course of a chemical reaction coupled with machine functioning). The efficiency of energy transduction is

$$\eta = W_{ext}/Q = (W - \Delta W)/Q = \eta_{id} - \Delta W/Q, \tag{3.5}$$

where $\eta_{id} = W/Q \leq 1$ (η_{id} may be less than 1 due to friction) corresponds to the maximal value of the conversion factor related to the ideal case of perfect force–load fitting ($F(X) = F_{ext}$). We can calculate ΔW identifying futile work with the energy payment for information, i.e., assuming $\Delta W \cong \Delta E$. Following Gray's arguments [44], we calculate ΔE from Heisenberg's uncertainty principle, $\Delta E \geq hc/(4\Delta X)$. Substituting $\Delta X \cong hc/(4\Delta E) \cong hc/(4\Delta W)$ into (3.4), we obtain

$$\eta = \eta_{id} - \alpha, \quad \text{where} \quad \alpha = 0.5\, Q^{-1} * \sqrt{hc \sum_i |F_i|}. \qquad (3.6)$$

Obviously, the physical meaning would have only those values of parameter α that satisfy the conditions $\alpha < \eta_{id} \leq 1$.

To compare the results for macroscopic and microscopic systems, let us take the same values of the model parameters as were used in Gray's work [44]. For example, if the process consists of the expansion of one mole of gas, $Q \cong 4 * 10^4$ J and $F_i \cong 10^5$ N, and thus the information energy expenditure can be neglected in all estimations of thermodynamic characteristics. For this set of parameters, $\Delta W/Q \ll 1$, and thus $\eta \cong \eta_{id}$. A quite different situation arises for the machines of molecular dimensions. The estimations for machines operating at the level of individual macromolecule parameters gives $\alpha > 1$. For instance, during the functioning of one myosin bridge, the hydrolysis of one ATP molecule produces $Q \cong 7 * 10^{-20}$ J; according to [45–47], the force created by one myosin bridge $F \cong 10^{-12}$ N. These values lead to $\alpha \approx 3$. Thus, for any reasonable values, $\eta_{id} \leq 1$, the obtained value of α leads to the absurd result $\eta < 0$.

The calculations presented above demonstrate that in order to ensure an equilibrium mode of molecular machine operation the cost of information appears to be too expensive. If the method for the evaluation of ΔE is valid, we have to conclude that:

(i) The use of light quanta is not an optimal method for obtaining the information that is necessary to perform the feedback control on the molecular level.

(ii) Optimal functioning of molecular machines becomes possible under the nonequilibrium conditions. In this case a machine does not need the information for performing feedback control.

3.2.3. Models for Calculating the Conversion Factor

In this section, we consider simplified models simulating the behavior of energy-transducing machines capable of performing mechanical work. These models allow us to compare the efficiencies of molecular machines operating under equilibrium and nonequilibrium conditions, demonstrating the advantage of the nonequilibrium mode of operation. We consider the behavior of an electromechanical machine consisting of several rigid elements (levers, hinges, pivots, etc.) that are able to change their mutual position under the external load or electric forces arising after the charging/discharging of cer-

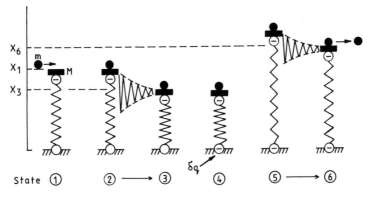

Fig. 3.2. Sketch of an electrochemical machine performing work in the equilibrium mode of operation.

tain elements of the construction. This machine also contains an elastic element as an essential part of its construction (for instance, a spring or elastic helix) that can change the length, being compressed under external force or extended due to the repulsing of electric charges bound to certain elements of the construction (Fig. 3.2). We have no intention of relating this model to any concrete biological process. Nevertheless, the analysis of the behavior of this simplest electromechanical device might be of some interest for simulating mechanical work performed during muscle contraction.

For the sake of simplicity, we consider the system having one mechanical degree of freedom as denoted by the variable X along the vertical coordinate axis. This "device" (Fig. 3.2) consists of the mass M attached to the top end of the vertical light helix (spring) characterized by the elasticity coefficient k. The extent of the elongation (or compression) of the helix is proportional to the force applied, $\Delta x = F/k$. Being unloaded, this light helix has a length a (the distance between the opposite ends). Also, let b be the length of the "heavy" helix, i.e., $b = a - Mg/k$. The lower end of the helix is fixed, while the upper end, bearing electric charge q, is able to move. Mass M is a part of the device constantly attached to the mobile end of the helix, which transforms it into a "heavy helix." The zero point of the coordinate axis, $X = 0$, corresponds to the position of the upper end of the free helix, while the coordinate $X_1 = -Mg/k$ corresponds to the equilibrium initial state of the helix compressed under the intrinsic weight M, but without any additional load (Fig. 3.2, state 1). An expanding helix can perform work against an external force, F_{ex}, applied to the top end of this helix. If $F_{ex} = \text{const.}$, then useful work equals the increase in the potential energy of a lifted weight of mass $m = F_{ex}/g$, where g is the gravity constant. Loading the system with the external mass m, placed at the top end of the helix (Fig. 3.2, state 2), causes further compression of the helix. We assume that the loading is performed fast enough, i.e., it occurs within the time interval $0 < \tau_{load} \ll \tau_{rel}$, where τ_{rel} is the

characteristic time of a system relaxation to the new state of equilibrium. After the relaxation of the system (Fig. 3.2, transition: state 2 →
state 3), accompanied by the energy dissipation into heat in the course
of decaying oscillations of the elastic helix, the top end of the helix
will reach a new equilibrium position, characterized by the coordinate
$X_3 = -(M + m)g/k$.

According to the model, useful mechanical work (lifting the external load
m) is performed as the result of the helix extension caused by the repulsing of
unipolar charges fixed at opposite ends of the helix. We assume that one of
the electric charges, q, is constantly bound to the upper end, while another
charge of the same sign, δq, instantly appears at the opposite end, and thus
initiates the extension of the helix (Fig. 3.2, state 4). In real biological systems,
the "creation" of an electric charge bound to the protein globule can be
realized, for instance, as the result of the dissociation of ions from certain
groups of a protein molecule. The appearance of electric charges, δq, in the
vicinity of another charge, q, is equivalent to the input of certain energy, ΔQ,
from the external source.

Considering the equilibrium mode of the machine functioning, we assume
that the charge δq appears after the loaded helix reaches its equilibrium
position, corresponding to thermal equilibrium between the device and its
surroundings (Fig. 3.2, state 4, $X_4 = X_3$). The repulsing between electric
charges causes extension of the helix, accompanied by lifting of the load m
(state 5). Under equilibrium conditions, unloading the system is also performed after attaining the new equilibrium state of the stretched helix (Fig.
3.2, state 6). After the displacement of the weight m from the machine, and
following removal of one of the electric charges, δq, from the device, the
system relaxes to the initial state of equilibrium (state 1). Useful work is
$\Delta W = m(X_6 - X_1)$. Removal of the electric charge δq in state 6 is equivalent
to the output of energy from the machine. If this energy is not wasted in
heat, the net input of energy from the external source, as related to one cycle
of the machine's functioning, can be evaluated as $\Delta Q = \Delta Q_{in} - \Delta Q_{out}$.
Here, $\Delta Q_{in} = (q * \delta q)/[D(a + X_3)]$ and $\Delta Q_{out} = (q * \delta q)/[D(a + X_6)]$, where
$(a + X_3)$ and $(a + X_6)$ are the distances between charges in the initial and
final states, and D is the dielectric constant of the medium between the
charges.

In the frame of this model, using the law of energy conservation and taking
into account the conditions for the balance of forces in the states of equilibrium, it is easy to calculate the conversion factor $\eta = \Delta W/\Delta Q$, which reflects
the efficiency of energy transduction. Introducing the dimensionless parameter $\alpha = (X_6 - X_1)/b$, characterizing the useful uphill displacement of the load
m from the initial position, X_1, to the final one, X_6, and parameter $\varepsilon = (X_1 - X_3)/b = mg/kb$, characterizing relative compression of the helix under
the external load m (the physical meaning has only those values of parameter
ε that obey the obvious condition $\varepsilon < 1$), we obtain the following expression
for the conversion factor corresponding to the equilibrium mode of the

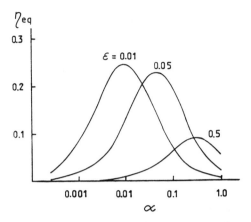

Fig. 3.3. Dependencies of the conversion factor, η_{eq}, on the relative extent of the spring stretching, α, for various values of the model parameter ε.

machine functioning:

$$\eta_{eq} = \frac{\alpha\varepsilon(1-\varepsilon)}{(1+\alpha)(\alpha+\varepsilon)^2}. \tag{3.7}$$

Figure 3.3 demonstrates how the conversion factor η_{eq} depends on parameter α for different values of parameter ε, which is determined by the intrinsic properties of the device (k and b) and the load (mass m), $\varepsilon = mg/kb$. These dependencies have a bell-like shape with the maximum values of η_{eq} determined by the parameter ε. Figure 3.4 demonstrates that the extreme values of

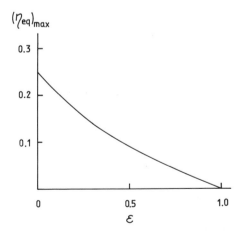

Fig. 3.4. Dependencies of the maximal conversion factor, $(\eta_{eq})_{max}$, on the model parameter ε.

the conversion factor $(\eta_{eq})_{max}$ do not exceed $(\eta_{eq})_{max} = 0.25$, corresponding to the limit $\varepsilon \to 0$. This means that it is impossible to reach a high enough efficiency of energy transduction with the equilibrium mode of machine functioning. Meanwhile, in real biological systems the energy transduction at the level of macromolecules occurs at higher efficiencies. For instance, using the experimental data concerning muscle contraction, the efficiency of performing mechanical work in the course of the filament's movement was evaluated as $\eta \cong 0.5$ [47].

Trivial calculations show that the efficiency could be increased if the machine could operate in the nonequilibrium mode. Speaking of the nonequilibrium mode, we mean that either loading or unloading of the device occurs at the machine cycle states preceding the establishment of thermal equilibrium. Optimal condition for machine operation should be realized in the following ways:

(i) the input of charge δq, initiating stretching of the helix, occurs immediately after loading the system, i.e., at the upper position of the helix, before relaxation to the equilibrium state of the loaded helix (Fig. 3.5, state 2); and

(ii) unloading the machine, i.e., the removal of the elevated load m, occurs at the upper position of the extended helix (Fig. 3.5, state 3) which precedes relaxation to the equilibrium state.

In the first case, when the input of energy occurs immediately after loading the system, but unloading is realized only after relaxation of the stretched helix to the state of equilibrium, trivial calculations lead to the following term for the conversion factor:

$$\eta^* = \frac{\varepsilon}{(1 + \alpha)(\varepsilon + \alpha)}. \tag{3.8}$$

Figure 3.6 shows a set of dependencies, $\eta^* = \eta(\alpha)$, calculated for various ε values. Comparing the data for two modes of machine functioning, we see that the nonequilibrium mode can provide higher efficiencies of energy conversion: $\eta^* \to 1$ in the limit $\alpha \to 0$, while $\eta_{eq} \to 0$ for $\alpha \to 0$ in the equilibrium mode. The physical meaning of higher efficiency in the nonequilibrium conditions of machine operation is quite clear: there is no useless dissipation of energy into heat that takes place in the course of the relaxation to the equilibrium after loading the system. The decrease in the η^* value with the enlargement of parameter α is, obviously, associated with increasing the amount of wasted energy during the relaxation of the extended helix: the longer the distance of the load movement, α, the greater is the amplitude of decaying oscillations (Fig. 3.2, transition state 5 → state 6). Another conclusion—the efficiency of the machine operation in the nonequilibrium conditions decreases with the enlargement of the helix elasticity (lower values of parameter $\varepsilon = mg/kb$ correspond to a higher elasticity coefficient k). In turn, this might also mean that for one and the same device ($k = $ const.) higher efficiencies η,

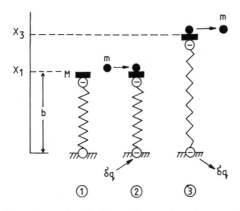

Fig. 3.5. Sketch of an electrochemical machine performing work in the nonequilibrium mode of operation.

corresponding to higher values ε, would be attainable in the case of a machine acting with the heavier load m.

Other nonequilibrium states arise immediately after stretching the helix (Fig. 3.5, state 3). Let us now calculate the conversion factor for our hypothetical device operating under the nonequilibrium conditions during two steps of the machine's cycle, after loading the machine and stretching a spring. In this mode of functioning, the withdrawal of the load m and further neutralization of δq should occur at the moment of time preceding relaxation of the extended helix (Fig. 3.5). Analytical treatment gives the following expression for the conversion factor:

$$\eta^{**} = \varphi^{-1} * [\sqrt{1 + 2\varphi} - 1], \tag{3.9}$$

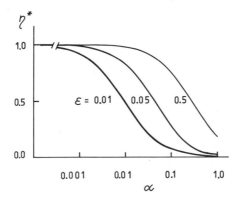

Fig. 3.6. Dependencies of the conversion factor η^* on the relative extent of the spring stretching, α, for various values of the model parameter ε.

where $\varphi = \Delta Q/\beta$ is a dimensionless parameter determining the efficiency of energy conversion for a nonequilibrium mode of the machine functioning. Here, ΔQ is the amount of energy inputted, $\beta = (mg)^2/k = F_{ext}^2/k$. Taking into account that $k = F_{ext}/\varepsilon b = mg/\varepsilon b$, we obtain $\varphi = \Delta Q/\beta = \Delta Q/\varepsilon b F_{ext}$. It follows from (3.9) that the maximal efficiency of energy transduction, $\eta^{**} \to 1$, can be attained with $\varphi \to 0$, corresponding to a "heavy" load ($F_{ext} \to \infty$), while $\eta^{**} \to 0$ in the limit $\varphi \to \infty$. The physical meaning of this result is quite clear: the replacement of a very light body ($F_{ext} \to 0$ and $\varphi \to \infty$) by moving a heavy one is inefficient.

Is it possible to reach high efficiency for this kind of machine having macromolecular sizes? Let us evaluate, for example, the range of parameter φ that might be reasonable for simulating real biological systems, for instance, a muscle's myofibrils. The force created in a muscle's sarcomer by one active myosin bridge has been evaluated as $F_{ext} \cong 10^{-12}$ N [45, 46]; the displacement of a myosin "head" is about the value $\Delta X \cong 10$ nm [47]. Taking $\Delta Q \cong 5 * 10^{-20}$ J (the standard change in Gibbs free energy of ATP hydrolysis) and assuming $\varepsilon b \cong \Delta X$, we obtain $\varphi = \Delta Q/(F_{ext}\Delta X) \cong 5$. According to (3.9), this value of parameter φ corresponds to $\eta \cong 0.46$ that is in consort with the conversion factor for the process of muscle contraction (about 49%) evaluated from experimental data [47].

Comparing factors of energy conversion for different modes of molecular machine functioning, we see that high efficiencies are attainable only for the nonequilibrium regime. To provide an optimal mode of operation the construction of the machine should contain a special device used for triggering the extension of the spring and further unloading the system at the right moments of time that precede the relaxation to the equilibrium states. The question arises: What is the energy cost for obtaining the information needed to gain the advantage of the nonequilibrium mode of machine operation? Let us evaluate the minimal portion of energy, ΔE, which must be paid in order to get adequate information about the position of the moving part of this hypothetical machine.

As a matter of fact, to provide a nonequilibrium mode of operation for our hypothetical machine we need to determine the position of the oscillating helix within the time interval Δt which is less than the period of oscillations, T, excited by loading the helix with the weight m (Fig. 3.2, state 2) or stretching the helix after the charge addition (Fig. 3.2, state 5). The period of oscillations in state 3 can be evaluated by the term $T = 2\pi\sqrt{(M + m)/k}$. According to quantum theory, $\Delta E \geq \hbar/\Delta t$, where ΔE is an uncertainty in the energy measured in time interval Δt, and $\hbar$ is the Planck constant. Thus, taking into account $k \cong mg/b\varepsilon$, we finally obtain

$$\Delta E \geq \frac{\hbar}{2\pi}\sqrt{\frac{mg}{(M + m)\varepsilon b}}. \tag{3.10}$$

For a "heavy load" model with the parameters in the intervals $\varepsilon b \cong 10$–100 Å, $M \cong 10^4$–10^6 in proton units and $m \cong 10^6$–10^7, formula (3) gives the value $\Delta E \cong (0.5-1.6) \cdot 10^{-30}$ J. Taking the energy input $\Delta Q \cong 5 \cdot 10^{-20}$ J (the standard change in the Gibbs free energy per one molecule of the ATP hydrolysis reaction), we obtain $\Delta E / \Delta Q \ll 1$. This result means that the energy payment for the information concerning the position of the moving part of the machine operating at a macromolecular level, evaluated according to the quantum mechanical rule of uncertainty, is very small as compared with the energy input for macromolecular machine functioning.

Comparing this result with Gray's treatment of a similar problem, we find that our evaluations give much smaller ΔE values, than that obtained by Gray [44]. This discrepancy can be explained. As a matter of fact, these two approaches to the ΔE evaluation imply using quite different frequencies for measuring distances or time intervals. Following Brillouin's treatment [39], Gray used formula $\Delta E \geq hc/4\Delta X$, where c is the velocity of light and ΔX is the uncertainty in ascertaining position X of the moving mechanical part (the piston in Gray's treatment). An application of this formula tacitly implies that the quantum of light should be used for more accurate measurement of the position of the moving part. What was the reason for using quantum of light to obtain the information on the right position of mobile parts of the machine? How would the controlling device, being an analogue of Maxwell's demon, measure the distances in the case of macromolecular machines? According to Brillouin [39], Maxwell's demon might measure Van der Waals forces or detect electric and magnetic fields generated by dipoles or magnetic moments. These forces are short-range ones. Therefore, the demon will react too late, and thus cannot provide "opening the door" at the right moments of time. Brillouin concluded that molecules can be detected using light long before they reach the wall, i.e., at the distances where we can neglect short-range forces. Measurement of the distances with high precision can be performed if $\hbar\omega/k_B T \gg 1$, where $\hbar$ is the Planck constant, k_B is the Boltzmann constant, ω is the circular frequency of light, and T is the temperature in degrees Kelvin. Using a "light-flash" for measuring distances is the reason why the amount of energy expended for obtaining the information appears to be too high at the molecular level. Electromagnetic irradiation for the measurement of distances is efficient if the wavelength is less than, or comparable with, the linear size of the object. This means that such a way of performing the feedback control needs illumination with a rather short wavelength, i.e., using the high energy quantum of light.

In Gray's approach to the problem of the reversible function of molecular machines [44], in order to reach best fitting between the load and the force created by the machine, the controlling device ought to monitor the position of a mobile part at any moment in time. To fulfill this program at molecular level, the controlling device would spend a relatively large amount of energy in obtaining the adequate information. Otherwise, *under nonequilibrium conditions of functioning, the machine does not need this excess of information.* What is necessary to provide the nonequilibrium mode of operation is only the knowledge of the right moments for triggering and loading (or unloading) the system. A relatively low-energy cost for this kind of information, that has been evaluated above for a simple putative model of an electromechanical machine, is a consequence of using the low-frequency (and thus low-energy) oscillations of a relatively heavy but mobile part of the device. In contrast to the former case, the physical state of the construction (*intrinsic conformational and dynamic properties of the device but not a quantum of short-wavelength light*) in itself provides the "instrument" for determination of the right moments for the loading/unloading of the machine. Thus, if it were possible to design such a more or less sophisticated device for triggering the loading/ unloading processes, then this controlling device would operate with a low-energy cost for information. We realize, however, that these kinds of analogies might be too far from real biological systems. The aim of our treatment was only to indicate, using a simple model, the advantage of the nonequilibrium conditions for functioning the energy-transducing machines. Gray and Gonda in their very interesting theoretical work [48], using a quantum mechanics approach, presented another line of argument as to why it is not necessary to expend energy for information if the molecular machine operates under nonequilibrium conditions.

The formal "electromechanical" model considered above illustrates how the nonequilibrium mode gains the advantage in energy conversion. The physical meaning of this result is clear: there is no wasting of energy into heat during relaxation to the equilibrium state at some steps of the machine cycle that take place under the equilibrium conditions of operation. In the frame of the models described, exciting the oscillations of the machine parts is equivalent to exciting the long-living mechanical degrees of freedom. In the course of the transition to a new equilibrium state the machine performs useful work. If the life-time of the nonequilibrium state is greater than the characteristic time of the energy-transducing act, then wasting energy into heat would be minimal. In other words, the storage of energy in the form of long-living nonequilibrium states (that also can be called "stored energy") provides high efficiency of energy transduction by the machines operating at the level of individual macromolecules.

3.3. Statistical Thermodynamics of Small Systems, Fluctuations, and the Violation of the Mass Action Law

As we noted in the preceding section, in spite of the microscopic dimensions of "molecular machines," each macromolecular device is large enough to be considered as a statistical system. Thus, the behavior of any individual molecular machine can be described in terms of statistical thermodynamics. On the other hand, there are certain "mechanical" features in the behavior of molecular machines. The reader can find a lot of theoretical and experimental data concerning various aspects of the structure and functioning of biopolymers in plenty of excellent monographs, and in original and reviewing articles (see, for references, [49]). The main processes of energy transduction in biological systems are realized by means of macromolecular complexes embedded in (or attached to) the membranes of closed vesicles of energy-transducing organelles (chloroplast's thylakoids, mitochondria, and chromatophores). Concerted functioning of these complexes (e.g., proton pumps driven by electron transport complexes and H^+ ATPases) ensures coupling between energy-donating and energy-accepting processes. The relatively small dimensions of energy-transducing vesicles impose, however, certain peculiarities on their functioning.

Our aim in this section is to consider some unusual thermodynamic properties of chemical systems that could reveal themselves by decreasing the system dimensions. For this purpose we will analyze a simple model simulating the chemical reaction $PQ \rightleftarrows P + Q$ within a small volume confined by a closed vesicle. We will see that small dimensions of vesicles under certain circumstances can impose limitations on the conventional thermodynamic description of chemical reactions based on the mass action law. The model considered may help to simulate chemical processes in the interior of small energy-transducing organelles.

3.3.1. Structural Peculiarities of Energy-Transducing Organelles of Chloroplasts

For an adequate quantitative description of thermodynamic relations inside the vesicles, associated with the processes of vectorial transmembrane transport of certain particles (hydrogen ions, Na^+ ions, etc.), we usually need to calculate the transmembrane difference in chemical potentials of these components, and the corresponding free energy changes accompanying a particle transfer from a vesicle to the surrounding medium, or the reverse. For many years, since Mitchell put forward, in 1961, his famous chemiosmotic hypothesis of membrane phosphorylation [50, 51], the measurement of transmembrane electrochemical gradients became the main touchstone for the justification (or refusal) of any concept concerning the mechanisms of membrane phosphorylation in energy-transducing organelles (see Chapter 5 for details).

Due to the small dimensions of energy-transducing organelles, the average number of certain components inside the bulk phase of the vesicles may be very low. Trivial calculations show, for instance, that at a neutral value of intrathylakoid pH an average number, $\langle p \rangle$, of free (unbound) hydrogen ions in the bulk phase of a single thylakoid is less than 0.3 [52]. To evaluate this number, let us consider some structural features of thylakoids.

Thylakoids represent closed membrane vesicles with relatively small internal volume. Being fixed under normal physiological conditions, and then examined with an electron microscope, thylakoids expose themselves as flattened discs (see sketch on Fig. 3.7) with diameter $d \approx 5000$ Å and thickness $h \approx 100$–200 Å, that would correspond to the internal volume $V \approx (2$–$4) *$ 10^9 (Å)^3. The experimentally determined internal (osmotic) volume of flat-

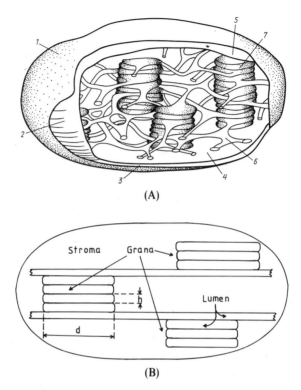

Fig. 3.7. Sketch of chloroplast (A) and its cross section (B). 1—external membrane, 2—internal membrane, 3—intermembrane space, 4—stroma, 5—grana, 6—stroma-exposed thylakoid, and 7—grana-exposed thylakoid.

tened thylakoids gives values of the same order of magnitude. Under normal isotonic conditions the osmotic volume of chloroplasts is $V_{osm} \approx 3$–9 μL per 1 mg of chlorophyll [53–55]. Taking into account that there are about three hundred electron transport chains per one thylakoid [56, 57], we obtain $V_{osm} \approx (1$–$3) * 10^9$ (Å)^3. For any reasonable intrathylakoid pH value, pH_{in}, a mean number of free (unbound) hydrogen ions, located in the bulk phase of relatively small intrathylakoid volume, appears to be insignificant. For instance, at $pH_{in} \cong 5$ ($[H^+]_{in} \cong 10^{-5}$ M), there is no more than six to twenty-four free H^+ ions per one thylakoid (parameter $\langle p \rangle$), while $\langle p \rangle \leq 0.3$ at $pH_{in} \cong 7$.

It should be stressed that the small sizes of vesicles in itself puts certain limitations on using conventional formula $pH_{in} = -\log_{10}[H^+]_{in}$. Perhaps, for such small vesicles as thylakoids, chromatophores, etc., we have no right to neglect the boundary effects that include the interactions of hydrogen ions, as well as other ions and molecules, with the internal surface of the vesicle wall. These interactions could cause deviations of the linear relation between the concentration and activity of ions inside a vesicle.

The effect of the membrane surface on the properties of an aqueous solution in the bulk phase of thylakoid lumen has been reported in [58, 59]. The microviscosity of the thylakoid's internal medium is at least ten times greater than that of the external medium. This may be a consequence of water "structuralizing," due to the effect of charged groups and proton exchangeable groups on the internal surface of the membrane. These groups, as well as soluble buffering groups, could influence proton diffusion, thus effecting the rate of the lateral transfer of H^+ ions inside the intrathylakoid bulk phase, and their traveling from the lumen of one thylakoid to another. The role of the boundary effects is determined by the fact that the characteristic thickness of a double electric layer near a charged membrane surface is comparable with the distance between the opposite internal surfaces of the thylakoid, $h \cong 100$ Å. So, the influence of a charged surface on an aqueous solution could spread over all bulk phases of the internal volume.

The problem of the adequate determination of internal pH is complicated in the heterogeneous populations of vesicles. Chloroplasts are a typical example of such systems. According to electron microscopy data, there are at least two morphologically different membrane systems: stroma- and grana-exposed thylakoids (see the sketch in Fig. 3.7). The numerous data in the last two decades brought evidence of the lateral heterogeneity of the chloroplasts' lamellar system (see references [60–69]). Stroma- and grana-exposed thylakoids differ with respect to the content of the electron transport and the ATP-synthase complexes.

Stroma-exposed thylakoids seem to be the continuation of grana-exposed thylakoids. There may, however, be some obstacles limiting the proton diffusion from the lumen of grana-exposed thylakoids into the lumen of stroma-exposed thylakoids. The hindrance to lateral proton transport may be due to kinetic reasons. Junge and his collaborators have come to the conclusion [70–72] that the apparent rate of proton diffusion within the lumen could be decreased by a few orders of magnitude lower than in the outer bulk phase. If R is the ratio of the number of protons bound to proton-exchangeable groups on the membrane, n_b, to the number of free hydrogen ions inside the vesicle, n_u, i.e., $R = n_b/n_u$, the apparent diffusion coefficient will decrease $(R + 1)$ times. Taking $R \approx 10^4$, Junge evaluated the apparent coefficient of lateral diffusion in thylakoid lumen as $D_H \approx 9 * 10^{-9}$ cm/s [73]. This value corresponds to the diffusion coefficient of the protein globule with the diameter $D \approx 47$ Å. Taking into account the peculiarities of two-dimensional diffusion, this diffusion rate will provide the proton displacement by the distance about 2000 Å for 10 ms (characteristic time of the chloroplast electron transport chain turnover). The distance ≈ 2000 Å is less than the average thylakoid in the lateral direction. Relatively slow lateral proton transfer could preclude randomizing the protons over the internal domains of stroma- and grana-exposed thylakoids, thus leading under certain conditions to the non-uniform distribution of proton chemical potentials between the thylakoids of both kinds (for experimental evidence, see Chapter 5).

The average number of free hydrogen ions is much less than the number of proton pumps driven by the chloroplast's electron transport chain. The greater part (99–99.9%) of protons pumped to the inside volume of thylakoids bind to buffering groups located on the membrane or in the lumen of thylakoids [52, 74]. Two questions arise:

(i) Is it correct to speak of the concentration of free hydrogen ions inside small vesicles if their average number can be as small as was calculated above (even less than one particle)?
(ii) What is the physical meaning of the pH value inside the vesicles of small sizes?

In Chapter 5 we will scrutinize the problem of experimental measurements of hydrogen ion concentration inside the thylakoids. We consider below some theoretical aspects of the problem.

3.3.2. Chemical Equilibrium Inside Small Vesicles

In accordance with the aforesaid, there may arise the problem of the correct calculation of chemical potential difference between inside and outside bulk phases of small vesicles. For macroscopic vesicles this problem is trivial, and can be easily solved using the well-known thermodynamic formula of the Gibbs–Nernst type for transmembrane difference in chemical potentials, $\Delta\mu_P = k_B T \ln(c_{in}/c_{out})$, where c_{in} and c_{out} are the concentrations of a component P inside and outside the vesicle, respectively. Under conditions of chemical equilibrium these concentrations can be calculated on the basis of the mass action law. For small enough vesicles, however, the problem of the adequate estimation of mean concentrations of particles inside a vesicle becomes more complicated. There are three reasons why the conventional thermodynamic approach to the calculation of particles concentrations could be misleading:

(i) enormous fluctuations in the number of particles within small vesicles;
(ii) the violation of mass action law; and
(iii) the discrete nature of particles leading to certain unusual peculiarities of the particle distribution over the whole ensemble of vesicles.

The conventional approach of equilibrium thermodynamics is applicable only in the case of the so-called thermodynamic limit: $N/V = $ const. at $V \to \infty$, where N is a number of particles in the volume V. This approach ignores all the factors mentioned above and being applied to small systems may therefore lead to erroneous results. For example, considering within the formalism of the thermodynamic approach the transmembrane transfer of neutral P particles from closed vesicles loaded with the reaction mixture of P and Q particles ($P + Q \rightleftarrows PQ$), we have obtained one paradoxical result: for small enough vesicles the transfer of even one P particle along its concentration gradient can be thermodynamically unfavorable under certain conditions

[75]. This result is, obviously, the direct consequence of formally using the conventional thermodynamic approach, neglecting the discrete spectrum of possible concentrations of the reagents confined within small volume.

In our further work performed in collaboration with Grosberg [76] we have considered the same hypothetical model system using the approach of equilibrium statistical mechanics that takes into account the discrete properties of the system. It was demonstrated that the puzzling result obtained in [75] is a consequence of two factors:

(i) the breakdown of the mass action law due to the fluctuations; and
(ii) "quantum" peculiarities of small systems having the discrete spectra of possible concentrations.

A similar problem of the apparent violation of the Second Law of Thermodynamics had faced Westerhoff and his colleagues, who used the nonequilibrium thermodynamics and kinetic approaches to analyze the problems of energy coupling in small vesicles [77–79] and the reaction yield in the so-called channeled systems [80, 81].

Model Description. Let us consider, following [75, 76], a simplified model system consisting of identical closed vesicles of inside volume V (Fig. 3.8). The vesicles and their surroundings contain neutral particles P and Q which participate in the reversible dissociation/association reaction, $PQ \rightleftarrows P + Q$. For the sake of simplicity we assume that in all other respects the system is ideal. We also suppose that the vesicle's envelope is impenetrable for P, Q, and PQ particles, at least for the time intervals sufficient to reach the dynamic equilibrium in the reaction $PQ \rightleftarrows P + Q$. The state of thermodynamic equilibrium is defined by giving the following parameters: the total number of particles P and Q inside the vesicle (parameters P and Q which include the numbers of free as well as of associated, particles in PQ form), a vesicle volume V, and equilibrium constant K.

According to conventional chemical thermodynamics, the equilibrium constant K is determined from the "*macroscopic*" mass action law, $K = ([P][Q])/[PQ]$, where $[PQ]$, $[P]$, and $[Q]$ are the concentrations of corresponding particles. "*Microscopic*" interpretation implies that the equilibrium constant K is the characteristic of a single PQ molecule, $K = v_0^{-1} \exp(\Delta E/k_B T)$,

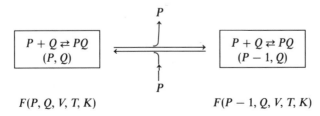

Fig. 3.8. See details in the text.

where ΔE is the energy difference between PQ and $(P + Q)$ states, and v_0 is the entropic factor of volume dimension. The value of the equilibrium constant K determines the average distribution of P and Q particles between their free and associated forms. For low values of K the equilibrium in the reaction $PQ \rightleftarrows P + Q$ is shifted to the left side; therefore, thermodynamically averaged numbers of free particles P and Q may be extremely small, in spite of the large total numbers of these particles inside the vesicle. In this special case, the chemical systems confined within a small enough vesicle will reveal some unusual thermodynamic properties that would be unnoticed in principle, if the concentrations are calculated with the conventional approach to chemical thermodynamics.

Statistical Treatment. Let us now calculate the change in the Helmholtz free energy, ΔF, that would take place after the unidirectional transfer of one particle P from the vesicle to outside. The choice of Helmholtz free energy as the characteristic function is determined by fixing two macroscopic parameters of our model system—temperature, T, and volume, V. By definition, the ΔF value is equal to the chemical potential of the free particles P inside the vesicle, $\mu_P = (\delta F/\delta P)_{V,T,Q}$. Since $\Delta F = \mu_P \Delta P$, $\Delta F = \mu_P$ at $\Delta P = 1$. This value can be calculated as the difference in free energies between two identical vesicles, one of which contains P and Q particles and another $(P - 1)$ and Q particles P and Q, respectively (Fig. 3.8). The correct calculation of the ΔF value, that takes into account the discrete nature of a system considered, can be performed on the basis of the equilibrium statistical mechanics approach. The same result is obtained from an alternative kinetic approach based on using a set of stochastic equations (for details, see [76]). Of course, there may be a great number of vesicles in a whole system, but all of them can function independently. Therefore, the evaluation of ΔF just for a single vesicle will have a physical meaning. The question of averaging over the ensemble of vesicles will also be considered below.

Let the integer p be the dynamic variable of the system that corresponds to the number of free (unbound) P particles inside the vesicle. For every fixed p value the system can be considered as the mixture of ideal gases with the numbers of particles p, q, and m; therefore, the partition function of our system is equal to

$$Z(V, K, Q, P) = \sum_{p=p_{min}}^{P} Z_p,$$

(3.11)

$$Z_p = (V^p/p!) * (V^q/q!) * [(VK)^m/m!].$$

Here, $q = (Q - P + p)$ and $m = P - p$ are the quantities of Q and PQ particles, respectively. The factor K^m corresponds to the energies of m bonds in the PQ molecule. For the sake of simplicity, we write down only the configurational part of the partition function, omitting here the terms corresponding to the so-called "quantum volume," arising from the integration over all generalized momentums, since this factor will drop out from the final expressions.

As was indicated by Sokirko (personal communication), it might be reasonable to consider the process $PQ \rightleftarrows P + Q$ in an aqueous medium, such that it does not change the total number of all particles occupying the volume V (for instance, dissociation of the water molecule, $2H_2O \rightleftarrows H_3^+O + OH^-$, does not change the total number of elementary cells occupied by all kinds of particles). In this case, we have to write down the expression for Z_p in the form that will differ from (3.11). However, trivial calculations demonstrate that for all interesting and nontrivial cases this model leads to practically the same results as our model for ideal gases approximation.

For the calculation of the observable $\langle p \rangle$ (corresponding to the averaged or experimentally measured mean value of variable p) and ΔF, we have following expressions

$$\langle p \rangle = \sum_{p=p_{min}}^{P} (pZ_p) \bigg/ \sum_{p=p_{min}}^{P} Z_p; \tag{3.12}$$

$$\exp(-\Delta F / k_B T) = Z(V, K, Q, P - 1) / [Z(V, K, Q, P)c_{out}]. \tag{3.13}$$

The c_{out} value determines the chemical potential of P particles in a macroscopic surrounding medium

$$\mu_{P_{out}} = k_B T \ln(c_{out}). \tag{3.14}$$

We want to emphasize that averaging in (3.12)–(3.13) relates to the number of free P particles but not to the concentration: $c_{in} = \langle p \rangle / V$. From (3.13)–(3.14) we obtain the Gibbs–Nernst equations for ΔF

$$\Delta F = -k_B T \ln(\langle p \rangle / V c_{out}) \equiv -k_B T \ln(c_{in}/c_{out}). \tag{3.15}$$

It follows from (3.15) that *the Gibbs–Nernst equation for the ΔF value is exact and holds true regardless of the system's volume V.* We have to emphasize, however, that the correct value of the equilibrium free P particles concentration inside the vesicle, $c_{in} = \langle p \rangle / V$, calculated within the framework of statistical mechanics, in general, may differ from that value formally calculated by the conventional thermodynamic approach, i.e., from the mass action law. This is the case for the reaction mixture of P, Q, and PQ confined within sufficiently small vesicle.

If the number of free P particles, p, is negligible in comparison with the total number of P and Q particles, i.e., if $p \ll P \cong Q$, we obtain the following expression for the average $\langle p \rangle$ value

$$\langle p \rangle = \sum_{p=p_{min}}^{P} p Y_p(n, \kappa) \bigg/ \sum_{p=p_{min}}^{P} Y_p(n, \kappa), \tag{3.16}$$

where $Y_p(n, \kappa) = \kappa^p / [p!(p-n)!]$, $n = P - Q$ and $\kappa = KVQ \cong KVP$. It follows from (3.16) that, in the nontrivial case of rather small values of parameter $\kappa \ll 1$ ($\kappa \to 0$ at $V \to 0$), the initial terms in (3.16) would play a dominant role in the determination of the $\langle p \rangle$ value. In the "macroscopic" case, corresponding to $\kappa \geq 1$, the dominant role in (3.16) will play the terms with higher p values. In this case, the p^* value, corresponding to the highest term, can be calculated using the conventional saddle point approximation, i.e., using the most probable term instead of the whole series [76]. It is not surprising that $p^* \cong (p)_{ma}$, where $(p)_{ma}$ is the number of P particles given by the mass action law.

Figure 3.9 demonstrates the numerical results of the $\langle p \rangle$ value versus the parameter $n = (P - Q)$ for a few values of parameter κ as were calculated according to (3.16), and the corresponding predictions of the mass action law (shown by continuous lines). The discrete nature of the system considered reveals itself as the explicit difference between the correct solution for the mean $\langle p \rangle$ value and the predictions of the mass action law. It follows from Fig. 3.9 that for the system confined by a sufficiently small volume V, the real mean number of free P particles $\langle p \rangle$ is essentially lower than the corresponding values, $(p)_{ma}$, predicted by the mass action law. On the other hand, for large enough ("macroscopic") systems, the mass action law gives practi-

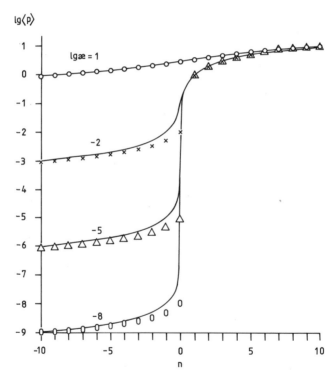

Fig. 3.9. Mean number of free particles P within the vesicle, $\langle p \rangle$, versus n, the difference between the total number of P and Q particles confined to the vesicle. The symbols are the result of statistical calculations, whereas the solid lines correspond to the predictions of the mass action law.

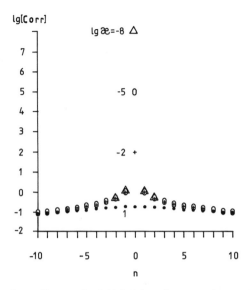

Fig. 3.10. The correlator, Corr $= \langle pq \rangle / \langle p \rangle \langle q \rangle - 1$, versus the $n = P - Q$ values for different values of the model parameter κ.

cally the same equilibrium number of P as that value calculated from statistical mechanics, $(p)_{ma} \cong \langle p \rangle$.

The most simple qualitative physical interpretation of this fact is the following. For low enough K values the decay of the PQ particle has a low probability, but if it takes place, the "newborn" particle P would spend a lot of time before it could meet a proper partner, a Q particle, as the result of a random walk in a large volume, and associate with it again. Otherwise, the probability of meeting the partner would increase by decreasing the volume V. Being confined within a sufficiently small volume, the "newborn" particles P and Q cannot go far away from each other, therefore the probability of their collision and association is higher than in a macroscopic system. We can say that the *free particles P and Q become correlated in a small volume* (Fig. 3.10). Therefore, the equilibrium concentration of free particles within a small volume, $\langle p \rangle / V$, is essentially smaller than that in a macroscopic mixture, $(p)_{ma} / V$.

In the "macroscopic" case of a large enough volume V, the evaluation of an equilibrium mean number of P particles, p^*, calculated from (3.16) using the saddle-point approximation [76], leads to the following relationship:

$$ K = \frac{p^*(Q - P + p^*)}{(P - p^*)V}. \tag{3.17} $$

We can see that (3.17) is the conventional expression of the mass action law for the reaction $PQ \rightleftarrows P + Q$. This is not the case, however, for sufficiently

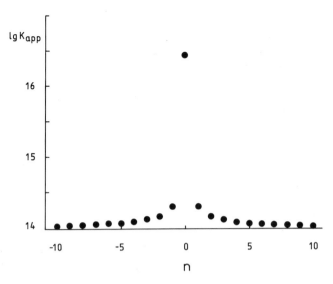

Fig. 3.11. An apparent equilibrium constant K_{app} versus $n = P - Q$ for the model parameters $\kappa = 6$ and $K = 10^{14}$.

small systems when significant fluctuations of the reagent concentrations lead to the violation of the mass action law. This law holds true only if the fluctuation are negligible. The analysis of formulas (3.12)–(3.15), as well as the numerical calculations (Fig. 3.9), demonstrate that for small values of parameter κ ($\kappa \ll 1$) significant fluctuations can lead to the essential violation of the mass action law. It follows from Fig. 3.11 that an apparent equilibrium constant K_{app}, calculated as (3.18) for true mean numbers of P, Q, and PQ particles, $\langle p \rangle$, $[Q - P + \langle p \rangle]$ and $[P - \langle p \rangle]$, may significantly differ from the K value corresponding to the "macroscopic" case (3.17)

$$K_{app} = \frac{\langle p \rangle [Q - P + \langle p \rangle]}{[P - \langle p \rangle] V}. \tag{3.18}$$

As was noted above, the model parameter K is the true constant whose value is determined by the molecular properties of the system. Since constant K is the characteristic of a single PQ molecule, its value does not depend on the volume of a whole system. On the other hand, the experimentally determined conventional "equilibrium constant" K_{app}, in the general case, is not a real constant. This constant, being calculated from the measured experimental concentrations of particles, will depend on the volume confining these particles. In the thermodynamic limit only, when we can neglect the fluctuations, the apparent equilibrium constant would be equal to the real one, $K_{app} \cong K = \text{const}$.

Ensemble of Vesicles. Figures 3.9–3.11 demonstrate that the thermodynamic behavior of the reaction mixture inside a single vesicle depends essentially on the parameter $n = P - Q$. For any fixed Q, the $\langle p \rangle$ values would depend on the total number of P particles within a vesicle. Let us now consider *an ensemble of identical vesicles* which can differ from each other with respect to the content of P particles, while all vesicles are being equilibrated with a medium characterized by the fixed chemical potential of P particles. This may be the typical case for the population of vesicles in a real biochemical preparation. Let the vesicles be impenetrable for Q and PQ, and let their concentrations inside the vesicles be held fixed. If the vesicles' walls are semi-transparent for P particles, then there may be an inhomogeneity with respect to the content of P particles inside various vesicles. If the ensemble of vesicles is incubated in the same medium with a certain concentration of P particles (or, say, at a fixed chemical potential μ_P), the number of P particles within various vesicles cannot be controlled with any finite accuracy. Statistical mechanics allows us to find the partition function, $W_P(V, K, Q, \mu_P)$, corresponding to the distribution of vesicles with respect to P particles.

$$W_P(V, K, Q, \mu_P) = [Z(V, K, Q, P)/\zeta(V, K, Q, \mu_P)] \exp(-P\mu_P/k_B T);$$

$$\zeta(V, K, Q, \mu_P) = \sum_{P=0}^{\infty} \exp(-P\mu_P/k_B T)Z(V, K, Q, P). \tag{3.19}$$

The numerical examples presented in Fig. 3.12 demonstrate that in the "macroscopic" case the W_P distribution is practically Gaussian. Meantime, for the "microscopic system" there may be an essential deviation of the distribution from the Gaussian one. The physical reason for such a deviation is, of course, the effect of fluctuations. The knowledge of the distribution function W_P is important, since it allows the calculation of average values of the different characteristics of the whole system, for example, such as the probability of finding a vesicle with a positive ΔF value, average free energy ($\overline{\Delta F}$), and so on.

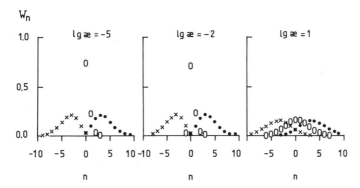

Fig. 3.12. A probability distribution for vesicles over different numbers of P particles, $P = n + Q$ (where $Q = $ const.) for three different values of the chemical potential of P particles in the surrounding preparation medium (see [76]).

If we consider a macroscopic system consisting of a multitude of independently functioning vesicles, an average free energy change, $\overline{\Delta F}$, related to the whole ensemble of vesicles is

$$\overline{\Delta F} = \sum_{P=0}^{\infty} W(V, K, Q, \mu_P) \Delta F(V, K, Q, P) \cong \sum_n W_n(\kappa, \mu_P) \Delta F(\kappa, n), \quad (3.20)$$

where ΔF is the corresponding value for one vesicle, which can be calculated according to (3.15)–(3.16). Reformulating (3.20), we can write

$$\overline{\Delta F} = \sum_{\Delta F} W_{\Delta F}(\kappa, \mu_P) \Delta F, \quad (3.21)$$

where $W_{\Delta F}(\kappa, \mu_P)$ is the probability of finding the vesicle within a given ΔF value in the ensemble of vesicles with fixed parameter κ ($\kappa = KVQ$) and μ_P. Figure 3.13 demonstrates one of the examples of numerical calculations for the partition function $W_{\Delta F}$. It follows from Fig. 3.13 that an average value $\overline{\Delta F}$, that corresponds to the experimentally measured value, can dramatically differ from the most probable one.

The simple model described above demonstrates one very interesting property of an ensemble of identical small vesicles. Being prepared under the same conditions, the vesicles from the ensemble can differ in the content of P particles. The total number of P particles within each individual vesicle are the large values, $P \gg 1$, but even minimal difference with respect to these values for two vesicles denoted, say, by indexes "i" and "j," $\Delta P = P_i - P_j = 1$, can lead to dramatic differences in their thermodynamic behavior. It may be possible that an average number of free P particles inside the "i"-vesicle, $\langle p \rangle_i$, is higher than that in the outside medium, $\langle p \rangle_i > \langle p \rangle_j$. If this is the case, the transfer of a single P particle (e.g., a proton from the chloroplast's thylakoid) from the "i"-vesicle will be thermodynamically favorable. This kind of phenomenon, associated with the highly nonhomogeneous properties of individual vesicles, which belong to the same ensemble, should be taken

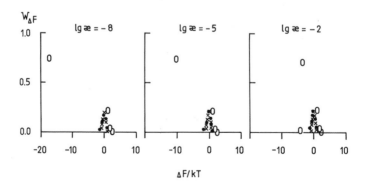

Fig. 3.13. A probability distribution for vesicles over different possible ΔF values of P particles, $P = n + Q$ (where $Q = $ const.), for three different values of the chemical potential of P particles in the surrounding preparation medium (see [76]).

into account for the explanation of certain experimental data concerning ATP synthesis by energy-transducing organelles (see, for references, [82–84]).

3.3.3. Compartmentalization and the Problem of the Macroscopic Description of "Channeled" Chemical Reactions

There are good reasons to believe that hydrogen ions relevant to membrane phosphorylation are nonuniformly distributed within energy-transducing organelles. In particular, under certain conditions the protons would not be completely delocalized inside the aqueous bulk phase of the intrathylakoid volume, but a significant portion of the protons would concentrate within certain membrane domains or special channels. There is also experimental evidence of the nonuniform distribution of hydrogen ions inside various kinds of thylakoids in chloroplasts. The activities of protons may differ significantly in the internal compartments of grana- and stroma-exposed thylakoids (experimental evidence will be considered in Sections 5.2.4 and 5.2.5).

On the other hand, the greater part of the usual experimental methods for measuring the concentrations of ions and their electrochemical potentials would obtain the information about these values without discrimination between various domains but averaged over all compartments containing the ions. These measurements can be done, for example, by determining the partition of special molecular probes between the internal volume of organelles (chloroplasts, mithochondria, etc.) and the outer bulk phase of the suspension [52–55]. Averaged thermodynamic parameters (ions activities, electrochemical potentials, etc.) could differ, of course, from the real thermodynamic parameters in certain functionally important compartments of the heterogeneous population of energy-transducing vesicles. This may be one of the reasons why an application of the nonlocal methods for the description of local events in the compartment isolated from others of a heterogeneous ensemble may lead to confusing results, and sometimes even to the *apparent* violation of the Second Law of Thermodynamics. Speaking of the nonlocal (or "macroscopic") approach to the description of thermodynamic parameters, we mean the methods that deal with the averaged values related to the whole ensemble under measurement. Some relevant examples from bioenergetics were scrutinized and analyzed in [77–83]. The problem of the adequate measurements of transmembrane proton gradients across the heterogeneous thylakoid membranes of chloroplasts will be considered in Section 5.2. Here we focus our attention on certain theoretical aspects of the problem.

Usually, an experimentalist calculates the chemical potential of a certain component P (e.g., hydrogen ions) by substituting the averaged concentration of this component, $\bar{p} = \overline{N}/V$, with the conventional formula for chemical thermodynamics

$$\mu_p = \mu_P^0 + RT \ln(\overline{N}/V),$$

where $\overline{N} = \sum_i n_i$ is the experimentally determined total amount of the component P accumulated inside all domains (or internal compartments of energy-transducing organelles) with the total volume $V = \sum_i v_i$. Here, n_i is the number of free P particles inside the compartment "i" of volume v_i. An experimentalist usually determines the total number of P particles, $\overline{N}$, confined to all compartments of the whole ensemble, as well as to the total internal volume V. Evidently, the averaged concentration, $\bar{p} = \overline{N}/V$, calculated from the measured values $\overline{N}$ and V, may be irrelevant to the real local concentrations inside the certain functionally important compartment "j," $p_j = n_j/v_j$, since the distribution of P particles over all domains, in general, is not uniform.

Parenthetically, we want to emphasize that the ergodic hypothesis, according to which the mean value obtained from averaging over the ensemble of particles is equal to the time-averaged value for an individual particle, does not in general hold true. Actually, let there be a large *kinetically equilibrium* ensemble of subsystems (e.g., vesicles) whose *interiors do not communicate* with each other. The interior of each individual vesicle from this ensemble is assumed to be under the equilibrium conditions, while the average concentrations of P particles inside various vesicles could differ significantly. For this reason, the time-averaged concentration of P molecules inside certain individual vesicle (denoted, say, by the index "j"), $\langle p_j \rangle$, can be essentially different as compared with the mean concentration $\bar{p} = \overline{N}/V$, calculated as an average concentration of P molecules related to the whole ensemble of vesicles. This would be true, at least, for any reasonably long period of time for which the time-averaging procedure could be carried out. The time interval for such averaging will be limited, of course, by the characteristic time of the particle exchange between the interiors of various vesicles of the ensemble. Usually, for P particles, the time of intervesicle exchange is much greater than the relaxation time of reaching the equilibrium states within each individual vesicle of the ensemble. For instance, the characteristic times of passive proton diffusion through the thylakoid membrane are $\tau \approx 5\text{--}10$ s at room temperature. The establishment of protonic equilibrium in the bulk phase of the intrathylakoid volume occurs with rates several orders of magnitude greater than that of the passive efflux of protons, i.e., practically there is no proton exchange between the thylakoids. Thus, the protons which appear inside the thylakoids would not have time to be equilibrated over the whole ensemble of vesicles. At least in time intervals less than τ, there would not take place the "randomization" of protons over the interiors of the whole ensemble of thylakoids. Therefore, for this characteristic time-interval, τ, the local concentrations, p_i, can differ rather significantly, as well as from the averaged (nonlocal or "macroscopic") value of the concentration, $\bar{p}$. These differences, however, may be masked if we use the conventional macroscopic description of the system.

The logarithmic dependence of the chemical potential μ_P of the component P on its concentration may drastically mask significant changes in the energetic characteristics related to a certain compartment "j," if chemical potentials are formally calculated on the basis of the experimentally measured averaged concentration of this component, $\bar{p} = \overline{N}/V$. To illustrate this state-

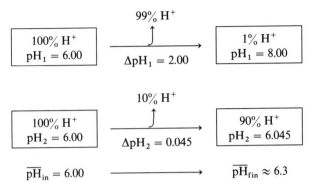

Fig. 3.14. See details in the text.

ment, let us consider the population of the vesicles of two kinds with equal internal volumes v. We assume that all vesicles initially contain equal numbers of free hydrogen ions, $n_1 = n_2 = n$, corresponding to an "average" internal $\overline{pH}_{in} = -\log_{10} \overline{p}_{in} = -\log_{10}(n/v)$. Let us assume that in the course of a certain process the concentration of H^+ inside the first kind of vesicle (50% of the whole ensemble of vesicles) decreases one hundred times, $p_1 = 10^{-2}n/v$, while for the second kind of vesicles the concentration decreases insignificantly, $p_2 = 0.9n/v$ (Fig. 3.14). Such a difference may occur due to an unequal leak of protons from the vesicles of both kinds. In the case of chloroplasts, for instance, the difference in proton effluxes from the thylakoids of grana and stroma is explained by the different content of the functioning ATPsynthase complexes [60–69]. The fast efflux of protons from the vesicles of the first kind through a large number of actively functioning ATPsynthase complexes could lead to a drastic increase in their internal pH_1, while, due to the low content of these complexes in the thylakoids of the second kind, the pH_2 value would increase, but not significantly. The final proton concentrations $p_1 = 10^{-2} n/v$ and $p_2 = 0.9 n/v$ correspond to the following "local" increases in the pH values: $\Delta pH_1 = 2$ and $\Delta pH_2 \cong 0.045$. Measuring the total amount of H^+ ions accumulated inside all vesicles, an experimentalist should conclude that there occurs a rather small increase in the "average" pH value, $\overline{\Delta pH} \cong 0.3$. Thus, using an averaged value of internal pH would mask a significant rise in the pH_1 value ($\Delta pH_1 = 2$) inside the large portion of vesicles. The functioning of these vesicles, however, might be responsible for the high biochemical activity manifested by the whole suspension of vesicles (e.g., ATP synthesis coupled with the proton efflux through functioning ATPsynthase). This example clearly demonstrates how the conventional phenomenological approach, based on using concentrations of the intermediates averaged over a whole system, could lead to false conclusions concerning the evaluation of local chemical potentials.

Analyzing the models for so-called channeled metabolic pathways, where a substrate S transformation into product P takes place inside separated compartments (or "channels"), Westerhoff and Kamp [80] have pointed to

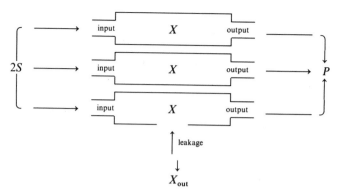

Fig. 3.15. See details in the text.

another reason for the essential difference between "deterministically" calculated and the actual rate of the reaction yield per channel. As a matter of fact, this reason is similar to that considered in the preceding section: the fluctuations in the number of certain intermediates inside small functional compartments. Let us consider this problem following the line of argument presented in [80]. Bimolecular reaction $2S \overset{k}{\to} P$ is catalyzed inside special "channels" (Fig. 3.15). According to the conventional ("deterministic" or nonlocal) approach, the rate of bimolecular reaction is

$$J_{det} = k(\bar{n})^2, \qquad (3.22)$$

where $\bar{n}$ is the average number of reacting S molecules per channel. On the other hand, according to the stochastic approach [85–87], the probability of binary collisions is proportional to $n(n-1)$, where n is the number of S molecules per channel (each of the n molecules S has a chance to meet $(n-1)$ other S molecules S). Thus, the actual rate of the binary reaction, J_{st}, should be expressed as

$$J_{st} = kn(n-1). \qquad (3.23)$$

The observed rate of chemical transformation, $\bar{J}_{st}$, relates to the rate value averaged over all compartments of the ensemble

$$\bar{J}_{st} = \sum_{n=0}^{\infty} [W(n)kn(n-1)] = k[\overline{(n^2)} - \bar{n}] = J_{det} + k(\sigma^2 - \bar{n}), \quad (3.24)$$

where $W(n)$ is the partition function which determines the probability of finding n molecules S inside the channel, and $\sigma^2 = \overline{(n^2)} - (\bar{n})^2$ is the variance related to the number of S molecules in the channel. For small enough compartments, the fluctuations in the number of reacting particles inside each individual channel may be rather large, and thus the variance σ^2 could reach significant values leading to the essential difference between the J_{det} and J_{st} values (for an illustration, see Fig. 3.16 reproduced from [80]).

We should remember these circumstances in order to be able to explain experimentally the observed thermodynamic "anomalies," including the puz-

Fig. 3.16. Dependencies of the reaction rate as calculated by the use of a stochastic (solid line) or deterministic approach (broken line). The figure is redrawn from [80], see details in the text.

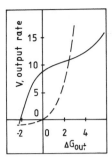

zling cases of the apparent violations of the Second Law of Thermodynamics (see Section 3.3.4). The deviation of the stochastic description from the deterministic one becomes, however, insignificant in the conventional (unchanneled) systems. In these systems all compartments are joined, thus forming one common reactor with a rather large number of molecules $\overline{N} = \overline{n}M$, where M is the total number of channels in the system. In this case, $\sigma^2 \to 0$, since we can neglect the fluctuations, and therefore $J_{det} \cong \overline{J}_{st}$.

Thus, if the material exchange by reacting particles between different compartments of the ensemble occurs more slowly than the establishment of the equilibrium (or steady state) within each individual compartment, the physical subdivision of the system into small and functionally separated compartments can be the reason for the divergency between the results given by the stochastic and deterministic approaches. Therefore, an application of the deterministic (nonlocal) approach might be inadequate for the description of the force–flux relations. Problems of this sort were analyzed at some length by several authors [82] who proposed a so-called "mosaic" coupling model for membrane phosphorylation. According to this and similar models, the protons transferred into a local compartment by its individual protonic pump are not equilibrated with other compartments, and thus cannot be competent in driving ATPsynthases located in other intracellular compartments or membrane domains. The mosaic model for membrane phosphorylation is very much like the so-called metabolic "channeling" of biochemial reactions [80, 81, 84].

3.3.4. The Fluctuations, Random Noise, Energy Transduction, and Apparent Violation of the Second Law of Thermodynamics

The conventional thermodynamic approach neglects the fluctuations, implying that the properties of the external medium are kept fixed with high precision, while the internal fluctuations are too low to be taken into account. As we have seen in the preceding sections, the fluctuations of thermodynamic functions in the interior of sufficiently small systems could be high enough. Now we want to focus our attention on the possible role of fluctuations in the behavior of biological energy-transducing systems.

The role of stochastic processes in the organization and functioning of biological systems was determined by many scientists working in the various fields of physics, chemistry, biophysics, and biochemistry [21–24, 77, 79–81, 85–91]. Even the properties of the rather simple macroscopic systems have appeared to be less "deterministic," as could be expected. For example, the dynamic system corresponding to the famous "strange attractor," described by the set of three nonlinear differential equations, reveals the turbulent behavior. This is an explicit example of how the purely deterministic system can generate "internal noise" (for reference, see [87–90]). A special interest represents the noise-induced transitions between the nonequilibrium states [89]. The mechanisms of self-organization in macroscopic systems associated with the formation of dissipative structures can lead to the appearance of rather complex structures [21, 23–24, 88–90]. If a "deterministic" system is kept far from equilibrium, there may appear a chain of bifurcations on the path of a system evolution. Therefore, the role of the internal and external fluctuations, inducing the transitions between various branches of nonequilibrium states lying in the path of the system evolution (or between different dissipative structures), becomes more significant than that for the system kept near the equilibrium.

In several theoretical models for small systems [77, 79, 80–81] the stochastic nature of the biochemical processes under certain circumstances reveals itself as the apparent violation of the Second Law of Thermodynamics. We consider below some of these models, demonstrating that the description of such processes in small systems based on averaging (nonlocal) formalism can often be misleading. These models demonstrate several unusual thermodynamic and kinetic properties that contradict the conventional laws of chemical thermodynamics and kinetics, described as a rule in terms of the *average concentrations* of the reagents and chemical intermediates.

Westerhoff and Chen [77] have investigated the model for the energy-transducing system (Fig. 3.17) in which the intermediate particles (e.g., protons) are injected into a closed volume within which the number of intermediates (e.g., protons or other relevant components) per coupling unit may be relatively low. Realizing the role of large concentration fluctuations in sufficiently small domains, the authors have treated the problem using a stochastic approach instead of the conventional "deterministic" methods of chemical kinetics. They calculated the steady state flux–force relationship, $\bar{J}_p = f(\bar{c})$, where $\bar{J}_p$ (the rate of ATP synthesis) and $\bar{c} = \bar{N}/V$ (an average proton concen-

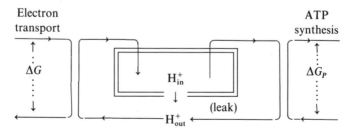

Fig. 3.17. See details in the text.

tration in the coupling box) are macroscopic variables. According to the model, the apparent driving force for ATP production, ΔG, was defined as $\Delta G = 2\Delta\bar{\lambda}_H - \Delta G_P$, where $\Delta\bar{\lambda}_H = RT\ln(\bar{c}/c_0)$, c_0 is the concentration of the intermediate in the surrounding medium, and ΔG_P is the phosphate potential. Both values, $\bar{J}_p$ and $\bar{c}$, represent the averages over an infinitely long period of time, that are assumed to be equivalent to averages over an infinitely large ensemble. This assumption is equivalent to ergodic hypothesis the correctness of which, however, for biopolymers and supramolecular complexes, can be questioned.

Westerhoff and Chen [77] were able to demonstrate that the rate of ATP synthesis (output flux) "is not a unique function of the average concentration of the intermediate when the latter occurs in small number." As the result of numerical simulation, under certain conditions the system displays the behavior (positive flux $\bar{J}_p$ generated by the negative driving force ΔG, Fig. 3.18) that contradicts the Second Law of Thermodynamics. This is the case when the volume of the domain in which the intermediate appears, V, is so small that an average number of intermediate $\bar{N} \leq 1$. Of course, for the macroscopic system the flux–force relationship follows the Second Law.

A similar result was obtained by Westerhoff and Kemp in [80] for the channeled system shown on Fig. 3.15. The rate of the output reaction versus the value $\Delta G_{out} = 2\mu_X - \mu_P$ (Fig. 3.16) reveals that positive flux (formation of P from two molecules of intermediate X) can be supported by the negative thermodynamic force ΔG_{out}. Chemical potentials of components X and P, $2\mu_X$ and μ_P, were calculated by using average concentrations of the reagents. This circumstance, i.e., using the averaged concentrations for the channeled system, is obviously the reason of the apparent violation of the Second Law.

Stochastic simulation has also demonstrated that it is impossible to find a single rate equation of the "deterministic" kind: the rate equation appears to be dependent on the *method* of variation of the number of particles inside the

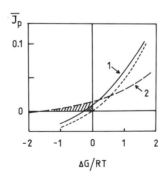

Fig. 3.18. Dependencies of the reaction rate as calculated with the use of a deterministic (solid line) or stochastic approach (broken line). The figure is redrawn from [77], see details in the text. Curves 1 and 2 correspond to different leaks in the model demonstrated in Fig. 3.17.

energy-transducing channel. According to the authors, the mean intermediate concentration "is not a useful thermodynamic variable" when this number in the box becomes small enough. They stressed, however, that the violation of the Second Law is only *apparent*, since the value $RT \ln(\bar{c}/c_0)$ cannot be identified as a true thermodynamic potential for small $\bar{N}$ values, or in the lack of the local equilibrium inside the energy-transducing domain. In these cases, the stochastic aspects of the process cannot be ignored.

The importance of the hydrogen ions concentration fluctuations inside small vesicles prompts us to consider the question as to whether or not the thermal fluctuations in the concentrations of certain intermediates (or of local electric potentials) could play any significant role in the energy-transducing processes. Significant potential fluctuations could occur in the vicinity of ion channels, $H^+ATPsynthase$ complexes, etc. These questions were also provoked by the very interesting experimental observations that an oscillating electric field can do chemical work being applied to the enzyme complexes. In particular, a regularly oscillating (1 kHz) electric field imposed on an erythrocyte suspension can drive the uphill transport of Rb^+ ions catalyzed by Na^+K^+–ATPase [92, 93].

The foregoing theoretical analysis carried out in [79] demonstrated that performing useful chemical work (or transport of substances) by "extracting free energy" from an electric field became possible if the processes catalyzed by an asymmetric enzyme complex were accompanied by a cyclic trans-membrane charge translocation. Theoretical calculations have also predicted that not only regular oscillations but totally random noise applied to a certain enzyme complex could produce chemical work. Stochastic simulation for the model systems in [79] suggested that random fluctuations of the transmembrane electric field generated by an *external* source can provide the uphill transport of substances against their concentration gradient. This should mean operating the *perpetuum mobile of the second kind*. What is a reason of such a result that evidently contradicts the Second Law, and what sort of noise might be competent in doing useful work by the molecular device extracting and converting free energy from the fluctuations in the surroundings? Common sense prompts us to state that randomly fluctuating

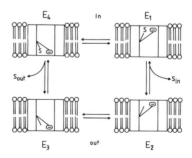

Fig. 3.19. Visualization of the model used in [79] (see details in the text).

local fields *in an equilibrium system* cannot be used for harvesting free energy through thermal noise. In this respect the authors of [79] questioned: "Whether the random noise used in our calculations is indeed characteristic for the noise generated by an equilibrium system in the vicinity of the enzyme?"

The reason for the apparent violation of the Second Law was indeed concealed in the unrealistic assumption concerning the nature of the random noise around the energy-transducing complex: at any moment in time the real *equilibrium noise* around the enzyme complex, *being affected by the enzyme state*, cannot be completely random. That is why internal noise cannot perform useful work if a system is kept under equilibrium conditions. An equilibrium noise in the immediate environment around an enzyme complex is *correlated* to the state of the enzyme. Taking into account the correlation between the enzyme state and the external ("autonomous") source of the noise produced by the surrounding, the authors [79] were able to solve the puzzle. Stochastic simulation for the system undergoing the action of the equilibrium (correlated) noise has led to the proper result, that *equilibrium noise cannot produce useful work*. Let us consider this model in more detail.

The numerical calculations were performed in [79] for the model system, simulating the functioning of a translocator embedded into a membrane (Fig. 3.19). The model considered potential-dependent steps in enzyme functioning: the movement of a negative charged arm of a protein across the membrane. The enzyme turnover cycle leads to the transmembrane transfer of the uncharged molecule. The fluctuations in the transmembrane electric field, $\Delta\varphi$, generated by an "autonomous" (noncorreleted) source of noise, can influence the charged protein-translocator. The effect of the translocator's states on the noise could be realized due to reciprocal electric interactions, e.g., as the result of the translocator's charge influence on the position of a noise-generating charged molecule. Thus, having charged elements of the construction, the translocator itself affects the noise in its surroundings. This correlated noise is really the equilibrium ("endogenous") noise. Monte Carlo calculations have demonstrated that there is no uphill transmembrane transfer if the reciprocal interaction between the translocator and generator of noise are taken into account. This example demonstrates that the equilibrium fluctuations in the electric field are strongly influenced by the translocator states. This can be compared with the famous Maxwell demon [39]. Being a device of molecular size, the demon would be strongly affected by the noise of the surroundings; therefore, a demon would fail to operate properly in picking out hot particles.

The authors of [79] concluded that "random electric noise generated by a free energy-dissipating process can do work.... As long as the probability of the onset of a fluctuation is independent on the enzyme state, work is done. However, equilibrium noise is strongly state-dependent and consequently does not lead to work."

We can find many other examples of the apparent perpetuum mobiles of the second kind. One of them is the Alkemade diode producing electric work by rectifying electric current fluctuations [87, 94]. Let the rectifier consist of the capacitor C and a nonlinear element (e.g., $n-p$ semiconductor contact) which are in a state of thermal equilibrium at temperature T. Due to the nonlinear Volt–Ampère characteristic of the element, $I = A *$ $\{\exp(eV/k_B T) - 1\}$, and the contact potential difference, there would appear an equilibrium charge, $\langle Q \rangle_e$, on the capacitor C. Conventional stochastic treatment based on the Langevin approach [87] leads to the value $\langle Q \rangle_e =$ $-0.5e$, where e is the electron charge. This result might mean that the device would rectify its own thermal fluctuations, thus supporting the voltage on the capacitor under equilibrium conditions. Being discharged through the load, the battery of such elements might perform work, thus violating the Second Law. This example illustrates how the macroscopic phenomenological approach becomes invalid for the analysis fluctuating nonlinear systems (for other examples and references, see [87]).

It is interesting to note that one unit of the "construction" (diode + capacitor) "generates" the equilibrium charge $\langle Q \rangle_e = -0.5e$ that is less than the elementary charge of a single electron. This means that electron transfer through the load, driven by the noise-generated potential difference $V =$ $e/2C$, cannot be favorable: a single step of electron transfer would create the potential difference of an opposite sign that, in turn, will "pull" the electron in the reverse direction. Therefore, to perform minimal work coupled with the transfer of one electron through the load, we need to coordinate the functioning of at least two units. However, it is not evident that fluctuating charges on both capacitors would appear as time-correlated (i.e., appear at the same moment of time) in order to provide the possibility of transferring one elementary charge through the load. Rigorous treatment of the problem has demonstrated that such a device cannot, of course, generate electric current by means of a unidirectional selection of the thermal fluctuations of charges in the equilibrium system [87].

References

1. L.A. Blumenfeld (1971), *Biophysics (USSR)* **16**, 724–727.
2. L.A. Blumenfeld (1981), *Problems of Biological Physics*, Springer-Verlag, Heidelberg.
3. C.W.F. McClare (1971), *J. Theoret. Biol.* **30**, 1–34.
4. C.W.F. McClare (1972), *J. Theoret. Biol.* **35**, 233–246.
5. C.W.F. McClare (1972), *J. Theoret. Biol.* **35**, 569–595.
6. C.W.F. McClare (1974), *Ann. N.Y. Acad. Sci.* **227**, 74–97.
7. A.B. Pippard (1957), *Elements of Classical Thermodynamics*, Cambridge University Press, Cambridge.
8. P. and T. Ehrenfest (1912), In: *Encyklopedie der matematischen Wissenshaften*, Band 4, Nr. 32, Teubner, Leipzig.
9. T.L. Hill (1960), *Statistical Thermodynamics*, Addison-Wesley, Reading, MA.

10. I.M. Lifshitz (1968), *J. Exper. and Theoret. Phys. (USSR)* **55**, 2408–2422.
11. *Models for Protein Dynamics* (1976) (H.J.C. Berendsen, Ed.), CECAM, University of Paris IX, France.
12. J.A. McCammon, B.R. Gelin, and M. Karplus (1977), *Nature (London)* **267**, 585–590.
13. M. Karplus and J.A. McCammon (1981), *CRC Crit. Rev. Biochem.* **9**, 293–349.
14. R.M. Levy, D. Perahia and M. Karplus (1982), *Proc. Nat. Acad. Sci. USA* **79**, 1346–1350.
15. M. Levitt and R. Sharon (1988), *Proc. Nat. Acad. Sci. USA* **85**, 7557–7561.
16. A. Warshel and S. T. Russel (1984), *Quart. Rev. Biophys.* **17**, 283–427.
17. A. Warshel (1984), *Proc. Nat. Acad. Sci. USA* **81**, 444–448.
18. L. Onsager (1931), *Phys. Rev.* **37**, 405–426.
19. L. Onsager (1931), *Phys. Rev.* **38**, 2265–2279.
20. I. Prigogine (1967), *Introduction to Thermodynamics of Irreversible Processes*, Wiley, New York.
21. P. Glansdorf and I. Prigogine (1971), *Thermodynamic Theory of Structure, Stability and Fluctuations*, Wiley–Interscience, London.
22. R. Balescu (1975), *Equilibrium and Non-equilibrium Statistical Mechanics*, Wiley–Interscience, New York.
23. G. Nicolis and I. Prigogine (1977), *Self-organization in Nonequilibrium Systems. From Dissipative Structure to Order Through Fluctuations*, Wiley, New York.
24. I. Prigogine (1980), *From Being to Becoming: Time and Complexity in the Physical Sciences*, W.H. Freeman, San Francisco.
25. A. Kachalsky and P.F. Curran (1965), *Non-Equilibium Thermodynamics in Biophysics*, Harvard University Press, Cambridge, MA.
26. T.L. Hill (1977), *Free energy Transduction in Biology*, Academic Press, New York.
27. T.L. Hill and E. Eisenberg (1981), *Quart. Rev. Biophys.* **14**, 463–551.
28. H.V. Westerhoff and K. van Dam (1979), In: *Current Topics in Bioenergetics* (D. Rao Sanadi, Ed.), Vol. 9, Academic Press, New York, pp. 1–62.
29. H.V. Westerhoff and K. van Dam (1987), *Thermodynamics and Control of Biological Energy Transduction*, Elsevier/North-Holland, Amsterdam.
30. S.R. Caplan and A. Essig (1983), *Bioenergetics and Linear Nonequilibrium Thermodynamics*, Harvard University Press, Cambridge, MA.
31. E.L. King and C. Altman (1956), *J. Chem. Phys.* **60**, 1375–1380.
32. R.M. Simmons and T.L. Hill (1976), *Nature* **263**, 615–618.
33. B.E. Banks (1969), *Chem. Brit.* **5**, 514–519.
34. B.E. Banks and C.A. Vernon (1970), *J. Theoret. Biol.* **29**, 301–326.
35. R.A. Ross and C.A. Vernon (1970), *Chem. Brit.* **6**, 539–540.
36. D. Wilkie (1970), *Chem. Brit.* **6**, 541–476.
37. T.L. Hill (1976), *Trends Biochem. Sci.* **2**, 204–207.
38. L.A. Blumenfeld (1983), *Physics of Bioenergetic Processes*, Springer-Verlag, Heidelberg.
39. L. Brillouin (1956), *Science and Information Theory*, Academic Press, New York.
40. R. Landauer (1961), *IBM J. Res. Dev.* **5**, 183–191.
41. C.H. Bennett (1982), *Int. J. Theoret. Phys.* **21**, 905–940.
42. R. Landauer (1985), *Ann. N.Y. Acad. Sci.* **426**, 161–170.
43. C.H. Bennett and R. Landauer (1985), *Sci. Amer.* **253**, 48–56.
44. B.F. Gray (1975), *Nature* **253**, 436–437.
45. C.R. Bagshaw (1982), *Muscle Contraction*, Chapman & Hall, London.

46. A.F. Huxley (1971), *Proc. Roy. Soc. London*, **B 178**, No. 1050, 1–27.
47. J.R. Bendall (1969), *Muscles, Molecules and Movement. An Assay in the Contraction of Muscles*, Heinemann, London.
48. B.F. Gray and I. Gonda (1977), *J. Theoret. Biol.* **69**, 167–186.
49. A. Fersht, *Enzyme Structure and Mechanism*, 2nd ed., Freeman, New York, 1985.
50. P. Mitchell (1961), *Nature* **191**, 144–148.
51. P. Mitchell (1966), *Chemiosmotic Coupling in Oxidative and Photosynthetic Phosphorylation*, Glynn Research, Bodmin, UK.
52. A.N. Tikhonov and L.A. Blumenfeld (1985), *Biophysics (USSR)* **30**, 527–537.
53. H. Rottenberg, T. Grunwald, and M. Avron (1971), *FEBS Lett.* **13**, 41–44.
54. S. Schuldiner, H. Rottenberg, and M. Avron (1972), *European J. Biochem.* **25**, 64–70.
55. H. Rottenberg (1979), *Meth. Enzymol.* **55**, 547–569.
56. H.T. Witt (1971), *Quart. Rev. Biophys.* **4**, 365–477.
57. H.T. Witt (1979), *Biochim. Biophys. Acta* **505**, 355–427.
58. D.M. Nesbitt and A.S. Berg (1982), *Biochim. Biophys. Acta* **679**, 169–174.
59. A.N. Tikhonov and A.A. Timoshin (1985), *Biol. Membranes (USSR)* **2**, 608–626.
60. L.A. Staehelin, P.A. Armond, and K.R. Millerr (1976), *Brookhaven Symp. Biol.* **28**, 278–315.
61. P.A. Armond, L.A. Staehelin, and C.J. Arntzen (1977), *J. Cell Biol.* **73**, 400–418.
62. K.R. Miller and R.A. Gusman (1979), *Biochim. Biophys. Acta* **546**, 481–497.
63. K.R. Miller (1980), *Biochim. Biophys. Acta* **592**, 143–152.
64. B. Andersson and J.M. Anderson (1980), *Biochim. Biophys. Acta* **593**, 427–440.
65. J.M. Anderson (1982), *FEBS Lett.* **138**, 62–66.
66. J.M. Anderson and W. Haehnel (1982), *FEBS Lett.* **146**, 13–17.
67. J.M. Anderson and R. Malkin (1982), *FEBS Lett.* **148**, 293–296.
68. J.M. Anderson and A. Melis (1983), *Proc. Nat. Acad. Sci. USA* **80**, 745–749.
69. W.S. Chow, C.Miller and J.M. Anderson (1991), *Biochim. Biophys. Acta* **1057**, 69–77.
70. A. Polle and W. Junge (1986), *Biochim. Biophys. Acta* **848**, 257–264.
71. W. Junge and A. Polle (1986), *Biochim. Biophys. Acta* **848**, 265–273.
72. W. Junge and S. McLaughlin (1987), *Biochim. Biophys. Acta* **890**, 1–5.
73. Y.Q. Hong and W. Junge (1983), *Biochim. Biophys. Acta* **722**, 197–208.
74. W. Junge, W. Auslander, A.J. McGeer, and T. Runge (1979), *Biochim. Biophys. Acta* **546**, 121–141.
75. A.N. Tikhonov and L.A. Blumenfeld (1990), *J. Phys. Chem. (USSR)* **64**, 1729–1740.
76. L.A. Blumenfeld, A.Yu. Grosberg, and A.N. Tikhonov (1991), *J. Chem. Phys.* **95**, 7541–7549.
77. H.V. Westerhoff and Y. Chen (1985), *Proc. Nat. Acad. Sci. USA* **82**, 3222–3226.
78. H.V. Westerhoff, T.Y. Tsong, P.B. Chock, Y. Chen, and R.D. Astumian (1986), *Proc. Nat. Acad. Sci. USA* **83**, 4734–4738.
79. R.D. Astumiam, P.B. Chock, T.Y. Tsong, Y. Chen, and H.V. Westerhoff (1987), *Proc. Nat. Acad. Sci. USA* **84**, 434–438.
80. H.V. Westerhoff and F. Kamp (1985), in: *The Organization of Cell Metabolism* (G.R. Welch and J.S. Clegg, Eds.) Plenum Press, New York, pp. 339–356.
81. F. Kamp and H.V. Westerhoff (1985), in: *The Organization of Cell Metabolism* (G.R. Welch and J.S. Clegg, Eds.) Plenum Press, New York, pp. 357–365.

82. H.T. Westerhof, B.A. Melandry, G. Venturoli, G.F. Azzone, and D.B. Kell (1984), *Biochim. Biophys. Acta* **768**, 257–292.
83. S.J. Ferguson (1985), *Biochim. Biophys. Acta* **811**, 47–95.
84. *Organized Multienzyme Systems* (G.R. Welch, Ed.), 1985, Academic Press, New York.
85. A.T. Bharucha-Reid (1960), *Elements of the Theory of Markov Processes and Their Applications*, McGraw-Hill, New York.
86. D.A. McQuarrie (1967), *Appl. Probab.* **8**, 1–66.
87. N.G. Van Kempen (1984), *Stochastic Processes in Physics and Chemistry*, North-Holland, Amsterdam.
88. H. Haken (1978), *Sinergetics*, Springer-Verlag, Berlin.
89. W. Horsthemke and R. Lefever (1984), *Noise-Induced Transitions*, Springer-Verlag, Berlin.
90. F.C. Moon (1987), *Chaotic Vibrations*, Wiley–Interscience, New York.
91. B. Gavish, in *The Fluctuating Enzyme*, (G.R. Welch, Ed.), Wiley, New York, 1986, pp. 262–339.
92. E.H. Serpersu and T.Y. Tsong (1983), *J. Membr. Biol.* **74**, 191–201.
93. E.H. Serpersu and T.Y. Tsong (1984), *J. Biol. Chem.* **259**, 7155–7162
94. C.T.J. Alkemade (1958), *Physica* **24**, 1029.

CHAPTER 4

Principles of Enzyme Catalysis

4.1. Introduction

In the preceding chapter we noted that the crucial elements of living matter function as chemical machines. The immediate sources of energy for all life activities are exoergonic chemical reactions. Certainly, the ultimate energy source for life on earth is solar light, but it can never be used directly without transforming the energy of light quanta into the energy of chemical bonds. Practically all important intracellular chemical reactions are accelerated by specific protein catalysts, i.e., enzymes.

What is a catalyst? According to textbooks on physical chemistry, a catalyst is a substance that encourages chemical transformations in other substances without itself being affected. This means that after each act of substrate transformation into a product the catalyst must return to its initial state. The initial and final states of an enzyme molecule do not alter, while the concentrations of substrate and product molecules (A_i or B_j in (2.1)) will change with the enzyme turnover.

Before going further in our analysis of the principles of enzyme action, we must say a few words about a widely held opinion among biochemists. Many of them think that there exists a general theory of catalysis in chemistry, and we need only to apply this theory to the more complex enzyme reactions. We think that this opinion is misleading. There is no such thing as a more-or-less self-consistent general theory of catalysis.

All catalytic chemical reactions can be subdivided into two types: homogeneous and heterogeneous. In the case of a homogeneous reaction, the catalyst is one of the low-molecular participants that is involved in the formation of an intermediate compound, thus facilitating the overall reaction to proceed on a new path along the reaction coordinate. There are numerous experimental data indicating that, in many cases, the enzyme reactions obey the same empirical (or semiempirical) rules as those governing the homogeneous catalytic reactions in solutions. When dealing with homogeneous catalysis chemists usually investigate the kinetics of chemical processes in the liquid phase. Actually, this branch of science is still in the process of the creation of a comprehensive theory.

It is rather doubtful, however, if enzyme catalysis can be considered as a kind of homogeneous catalysis. In this case, the catalyst is not just another low-molecular reactant, but a huge protein molecule with a defined surface that plays an essential role in the catalytic process. For this reason, many scientists relate enzyme catalysis to heterogeneous catalysis.

If we look at the experimental and theoretical aspects of heterogeneous catalysis, we will see that the state of affairs here is even worse than in enzyme catalysis. This is mainly due to the fact that it is easier to perform a reproducible experiment with purified enzymes than in the case of many solid catalysts. To be sure, enzymes are complicated substances, and it is necessary to be extremely careful during their preparatory isolation. However, being synthesized within the cell of a matrix, enzyme molecules represent practically identical copies. Therefore, the reproducibility of their functioning is, as a rule, much better than that of the solid catalysts of abiogenic origin. The identity of enzyme molecules is limited, however, by the possibility of the existence of "plural" forms of the enzymes, that are determined, in turn, by the chemical composition and the primary chemical structure (the sequence of amino acids residues in the polypeptide chain) of a protein molecule. Most proteins are sufficiently complex, and thus can exist in several conformational modifications.

The notion of enzyme activity requires certain elucidation. What is the meaning of the statements: "A chemical reaction is accelerated by the enzyme." With what should we compare the rate of enzyme reaction? Is this the rate of the same reaction running without the enzyme? As a matter of fact, "the same reaction" may be impossible without the enzyme. For most enzyme processes, there is a sequence of several reactions with several stable and unstable intermediates. Hence, it is not always possible to realize the same path of the overall reaction without the enzyme. Kosower introduced the concept of a so-called "*congruent model system*," i.e., *a chemical system in which the same overall reaction with the same stable intermediates is actualized* [1]. The reaction path between stable intermediates can of course be essentially different. We should therefore compare the rate of the enzyme reaction with the rate of a chemical transformation of a corresponding congruent system.

It is beyond our scope to describe here, in detail, the properties of enzymes, or all the factual data concerning the processes of enzyme catalysis. We can find this kind of information in numerous excellent textbooks on biochemistry (see, for instance, [2]). Although the reactions catalyzed by enzymes and chemical properties of various enzymes can be extremely different, we can always formulate certain specific features that distinguish enzyme catalysis from ordinary chemical catalysis. Some peculiarities of enzyme catalysis should be noted here.

The catalytic activity of certain enzymes (but not all) can be extremely high. The rate of an enzyme reaction can exceed that of a corresponding congruent model process by, for example, eighteen orders of magnitude. Enzyme catalysts are, as a rule, extremely specific. Apart from a few exceptions, an enzyme catalyzes only one definite chemical transformation from only one certain substrate. Let us consider, for instance, the enzymatic activity of iron ions and catalases. Catalase reaction is the decomposition of hydrogen peroxide into water and molecular oxygen

$$2H_2O_2 \rightleftarrows 2H_2O + O_2.$$

This reaction is an important biochemical process, and therefore catalase is widely spread over the different cells of all living organisms. Hydrogen peroxide, formed as a collateral product in many biochemical reactions, can be very harmful to intracellular components. Thus, catalase plays an important role as the biochemical protector preserving different compartments of a cell from a toxic agent, H_2O_2. Iron ions in water solutions reveal different kinds of catalytic activities: oxidase, peroxidase, and catalase. The activity of iron ions incorporated into catalase (containing four haems with iron ions) in performing the decomposition of hydrogen peroxide increases enormously. One catalase molecule destroys five million hydrogen peroxide molecules per minute. The activity of one milligram of catalase is equivalent to that of two kilograms of iron ions in water solution.

The temperature dependencies of catalytic activity are qualitatively similar for all enzymes (Fig. 4.1). For any enzyme, there exists a so-called "point of temperature optimum," i.e., the temperature of maximal catalytic activity. At the low-temperature branch of the temperature dependence, the increase in the enzyme activity with rising temperature is usually explained within the Arrhenius mechanism. This viewpoint, however, seems to be oversimplified. The temperature rise not only enhances the Boltzmann factor, increasing the probability of overcoming the potential barrier, but also influences the structural properties of an enzyme molecule. According to the majority of textbooks on biochemistry, the decrease in catalytic activity at high temperatures is due to the reversible inactivation of an enzyme or its denaturation occurring with a sufficient increase in the temperature. If the latter factor dominates, the value of the temperature optimum will be determined in prac-

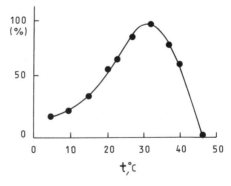

Fig. 4.1. The temperature dependence of the chloroplast photosystem 2 activity (after [61]).

tice not only by the particular enzyme properties, but also depends on the skill of the experimenter who has to handle the enzymes rather quickly to prevent damage to them at high temperatures.

The most simple general picture of enzyme catalysis is given by the scheme proposed by Michaelis and Menten

$$E + S \rightleftarrows ES \rightleftarrows EP \rightleftarrows E + P. \tag{4.1}$$

The substrate, S, and enzyme, E, form the so-called Michaelis complex, ES, the decomposition of which leads to substrate chemical transformation into the product, P. Let us assume, for the sake of convenience, that the substrate concentration, $[S]$, is essentially higher than that of the product, $[P]$. Consequently, we can neglect the backward reaction of the final stage in the scheme (4.1). The substrate attachment to the enzyme, as well as any local perturbation of the protein macromolecule (ligand association or dissociation, redox change of a metal ion in the active center, protonation or deprotonation of certain acid groups, etc.) lead, as a rule, to the formation of a protein specific nonequilibrium state. After such fast perturbation, the local vibrational relaxation proceeds extremely quickly (10^{-10}–10^{-12} s), but the relaxation of the whole protein globule to the new equilibrium state takes a longer time, up to a few hundred milliseconds. Thus, substrate binding to the enzyme catalytic center can lead to the formation of a rather long-living nonequilibrium state in which the immediate surroundings of the catalytic center are already changed, but the largest part of the protein macromolecule has practically the old conformation. References to numerous experimental papers concerning studies of such nonequilibrium states and processes of their relaxation can be found in [3, 4].

There are phenomena such as allosterism and the so-called *"enzyme memory"* that are closely related to the protein conformational changes. Allosteric enzymes change their catalytic activity if a certain ligand binds to the periphery of the protein globule far from its catalytic center. *"Mnemonic enzymes"* change their catalytic activity during their functioning. The initial activity of these enzymes is comparatively low but increases with time after many turnover acts (for references, see the two papers concerning hexokinase [5, 6]). This memory of past catalytic acts is connected with a comparatively slow enzyme conformational transition.

4.2. Earlier Theories of Enzyme Catalysis

Analyzing existing theoretical approaches to explain the mechanism of enzyme functioning, we can conclude that all points of view concerning the origin of the extremely high catalytic activity of enzymes may be subdivided into four groups.

  (i) The formation of the Michaelis complex increases the frequency factor in the Arrhenius equation (or S_a in the equation of activated state theory).

(ii) The formation of the Michaelis complex lowers the reaction activation energy, E_a.

(iii) The recuperation of energy liberated in the substrate–enzyme complex formation and using it to overcome the activation barrier.

(iv) The increase in the probability of specific thermal fluctuations in the Michaelis complex (as compared with the isolated substrate) leading to the excitement of certain vibrational degrees of freedom.

The first two explanations are the natural consequences of classical kinetics. According to transition-state theory, as generally accepted by biochemists, the value of a rate constant, $k = (k_B T/h) \exp(S_a/R) \exp(-E_a/RT)$, is determined by two main exponentially dependent factors. The rate constant can be increased either by increasing activation entropy or by decreasing activation energy.

Let us start our analysis with consideration of the first factor, i.e., with the "entropic explanation" of enzyme activity. This approach has been used by Koshland to describe those kinds of processes that, when drawing together substrate molecules, ions and substrates, cofactors and substrates, etc., are the obligatory intermediary steps determining the overall reaction rate [7, 8]. In the frame of the transition state theory, the configuration of the activated complex, that corresponds to the lowest potential barrier, is reached with the precise position and mutual orientation of the reactants. This means that such ordering of the reagents, in the course of an activated complex formation, would be accompanied by the initial entropy decrease, i.e., by the positive contribution to the activation entropy, S_a, of a model system.

The preliminary formation of Michaelis complex would begin with the necessity for the initial optimal orientation of substrate molecules (or substrate and cofactor particles, etc.) bound to an enzyme active center. In this case, movement along the reaction coordinate to the formation of the activated complex starts from the "proper" initial arrangement of interacting groups and molecules. Hence, the above-mentioned negative contribution to activation entropy, that should take place in the case of the activated complex formation in the congruent reaction, will be diminished in the enzymatic process. Due to this entropic factor the reaction rate should increase. Theoretically, the acceleration of the reaction due to this effect could be very high. The extent of the acceleration is determined by the range of angles between the reacting components which are postulated to be appropriate for providing the step of chemical transformation. The less the range of angles assumed, the greater will be the calculated acceleration. According to Koshland, the acceleration may reach eighteen orders of magnitude.

Milstein and Cohen [9] tried to verify experimentally Koshland's approach by comparing the rates of two kinds of chemical processes:

(i) the *bimolecular* acid-catalyzed esterification reaction between free acetic acid and phenol in the solution (Fig. 4.2(A)); and

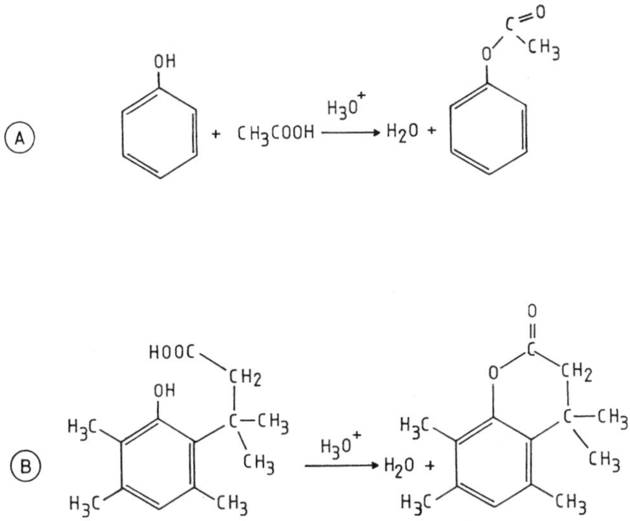

Fig. 4.2. Bimolecular (A) and monomolecular (B) esterification reactions considered in [9].

(ii) the *monomolecular* reaction between the same groups, acetic and hydroxy, which were bound together to the same aromatic ring.

The authors had investigated the kinetic characteristics of several derivatives that were different in the extent of their acetic group mobility (one of these reactions is shown in Fig. 4.2(B)). The rate constant of the bimolecular reaction (Fig. 4.2(A)), $k_A \approx 10^{-10}$ M^{-10} s^{-10}, and of the monomolecular reaction (Fig. 4.2(B)) reaches $k_B \approx 6 \cdot 10^5$ s^{-10}. At unit initial concentrations of the reactants, the rate of esterification with a fixed mutual arrangement of the reacting groups is thus greater, by sixteen orders of magnitude, than the rate of bimolecular esterification.

This result, however, cannot be considered as an unambiguous experimental confirmation of the possibility to accelerate enzyme reactions owing to the entropy factor. Rigorously speaking, the quantitative comparison of the rates of monomolecular and bimolecular processes, as related to equal concentrations of the reagents, seems to be incorrect. Under these conditions, the efficient "local" concentration of reacting groups bound to the same aromatic ring is always much greater than the concentrations of similar free groups in the solution. Taking into account the flexibility of bound groups and their mobility within the available but limited volume in the vicinity of the aromatic ring we obtain, for solutions with a 1 mM concentration of the reagents, that the probability of collision between two reacting groups attached to the organic ring will be about 10^5 times greater than that value related to the bimolecular reaction. The monomolecular reaction (Fig. 4.2(B)), that is to be compared with bimolecular reaction (Fig. 4.2(A)), is actually the reaction of a new, *previously synthesized* compound.

Concerning the enzyme catalysis, we must consider all its stages. The formation of an enzyme–substrate complex with a fixed mutual orientation of reagents must evidently be accompanied by a decrease in the system entropy of at least the same value, as the value of activation entropy increases. To achieve an increase in the reaction rate by eighteen orders of magnitude, as required by Koshland, the system entropy must first be diminished, due to formation of the Michaelis complex, by at least 83 e.u. Hence, to compensate for such an entropy drop at the formation stage of the enzyme–substrate complex, the substrate binding energy at room temperature cannot be less than a value of about 100 kJ/mole.

This "entropic" explanation of the enzyme functioning is based on the assumption that all the postulates of classical chemical kinetics are fulfilled. Substrate molecules undergo chemical transformation, and their relative movement along the reaction coordinate should be regarded as a thermal motion. The strong bond between enzyme and substrate, which effectively freezes translational and rotational degrees of freedom, must hinder the movement along the reaction coordinate and therefore increases the apparent energy of activation. Moreover, completion of the enzymatic process requires dissociating the reaction products and returning the enzyme molecule to its initial free state. It may happen that with increasing energy of the substrate–enzyme bond, the limiting step of the overall process will be dissociation of the enzyme–product complex. Certainly, part of the energy, liberated after the passage of the reacting system through the activated state, can be transformed into a new form *without dissipation*, thus resulting in diminishing energy of the enzyme–product bond. The question arises as to what the mechanism of this process is. In the realm of classical chemical kinetics, the specificity of substrate binding, leading to the unique mutual arrangement and orientation of reacting groups, cannot explain such a high extent of enzymatic activity stimulation, as compared with the rate of the same reaction in a congruent model system. Similar reasoning has been considered by Lumry [10].

According to the second theoretical approach, an enzyme lowers activation energy. This case is in agreement with the multitude of experimental data based on the measurements of the temperature dependence of enzyme reaction rates. As we have considered above (see Chapter 2), for enzymatic reactions the Arrhenius-like determination of activation energy is meaningless, and can thus lead to quite erroneous values and misleading conclusions. Such experimental data cannot therefore be considered as the justification of the statement that the formation of the Michaelis complex lowers an activation barrier, E_A. For this kind of explanation, all concrete schemes proposed can be reduced to the following trivial statement: the formation of a substrate–enzyme complex results in the decreasing transition-state energy,

E_a, by changing the reaction path. The critical analysis of certain concrete schemes concerning this concept can be found in [3].

A very important approach to the problem associated with the consideration of protein conformational changes had been put forward by Bauer as early as 1935. According to [11], protein molecules within a living cell exist in a unique "*steady nonequilibrium*" deformed state. In the course of an enzymatic reaction proteins undergo transition to the equilibrium state, and the energy liberated during this transition is used to overcome the activation barrier. After the passage of a substrate through the activated state its energy decreases; liberated energy does not dissipate but is used to return the protein to the initial strained state. This sort of explanation now seems too oversimplified and controversial from the energy viewpoint. However, having appeared more that 55 years ago, when no-one knew anything for certain about conformational changes of proteins, this idea was really surprising. The following generations of scientists could add but little to this hypothesis. The term "*energy recuperation*" relevant to this hypothesis was introduced by Kobosev [12] who did not know about Bauer's work (Bauer perished as a victim of Stalin's great terror in the late 1930s, and almost all issues of his book had been destroyed). According to Kobosev [12], energy recuperation means that a protein macromolecule ensures the process of the utilization of energy liberated in the course of catalytic reaction. This energy is used for the activation of an enzyme catalytic center. In an interesting but now forgotten book [13], Medvedev suggested that an enzyme molecule was able to store energy liberated in the course of catalyzed reaction, and then use it in the following catalytic act.

Further progress has been achieved in the work by Churgin et al. [14], who were the first to propose that the "normal" degrees of freedom of the macromolecule were separated from the "constructional" (mechanical) ones. The authors suggested a slow dissipation of energy initially localized on these mechanical degrees of freedom. Explicit interpretation of the general mechanism of enzyme catalysis based on the concept of energy recuperation has been given by Shnoll [15]:

"The conformation of an isolated enzyme molecule differs from that of a protein in the substrate–enzyme complex. The protein macromolecule is constructed *in such a way* that after completion of a step of substrate transformation (as the result of random thermal fluctuation) the energy liberated causes the transformation of the enzyme molecule into a special nonequilibrium state of thermodynamically unfavorable conformation. For kinetic reasons (large activation barrier) the macromolecule can remain in this state for quite a long time (in general, this conformation might be characterized by an extremely long life-time). The nonequilibrium macromolecule may be triggered into a new probable state due only to contact with a substrate. We have a peculiar but quite realistic paradox: the substrate catalyzes the enzyme transition into an equilibrium state. In turn, the substrate undergoes the necessary transformation, whose activation barrier diminishes at the expense of the *conformational* energy of the enzyme macromolecule." The most important feature of this version of the enzyme action is that the conformational changes are considered to be obligatory at all crucial steps of the catalyzed reaction. The theoretical foundation for this version, however, has a few essential shortcomings. The nature of the very long-living nonequilibrium state of free (without substrate) enzymes is rather vague. Also, it is difficult to imagine the physical mechanism explaining how the transition of a macromolecule into a conformationally nonequilibrium state can be induced by the local act of the substrate–product transformation. The authors state only that "the protein molecule is constructed in such a way that ...."

The fourth approach to the explanation of the enzyme functioning was proposed by Moelwyn-Hughes in 1959 [16]. This approach is based on the well-known work of Hinshelwood on the theory of unimolecular gaseous reactions at low pressures [17]. According to this work, the enzyme molecule increases the probability of useful energy fluctuation. Let us assume that the elementary act of chemical transformation, which determines the rate of overall process, is the dissociation of the ith bond in a multiatomic molecule. This stage will be actualized if the average energy of the vibrational degree of freedom, corresponding to exciting the stretching vibration of the ith bond, ε_i, is greater than (or equal to) the activation barrier of bond dissociation, E_a, i.e., $\varepsilon_i \geq E_a$. In the condensed phase the frequency of molecular collisions, spreading the energy through the whole system, should be rather high. This condition implies that the probability of energy accumulation from other vibrational degrees of freedom on this particular ith bond is very low.

If the gas pressure is low and, thus, the collisions frequency is not high, the above-described elementary act could take place at the less strong requirement, $\sum_i \varepsilon_i \geq E_a$, where summing corresponds to all vibrational degrees of freedom in the reacting molecule. At low collision frequencies, there is sufficient time for the "energized" molecule to use the fluctuation of vibrational energy. In this case, energy will be localized at one ith degree of freedom. This is equivalent to decreasing the measured activation barrier. Therefore, all molecules whose total vibrational energy is greater than or equal to the activation barrier can enter into the reaction. Obviously, this theory is based on the assumption that the excess of vibrational energy of a molecule may not have enough time to dissipate due to intermolecular collisions. Therefore, this theory would hold true only for gaseous reactions at low pressures. For enzymatic reactions in a condensed phase, this assumption cannot be fulfilled.

4.3. The Relaxation Concept of Enzyme Catalysis

The theories on the conformational changes of the enzyme molecule in the course of a catalyzed reaction and the crucial role of these changes for enzyme functioning were put forward many years ago. The history of this problem has been considered in brief in the preceding section of this chapter. The first approach toward the relaxation concept of enzyme catalysis was probably formulated in 1970 in the pioneering work by Sidorenko and Descherevsky. According to [18], after the completion of the enzymatic cycle and the liberation of a product, the free enzyme molecule remains in a conformationally nonequilibrium state. The relaxation time of this state may be comparable with the time interval between two consecutive catalytic acts related to a single enzyme molecule. In this case, catalytic activity of an enzyme will depend on the time moment of a substrate attachment to the enzyme molecule after a product dissociation in the preceding cycle of enzyme turnover. Two mathematical models have been analyzed. According to the first, "discrete" model, an enzyme can exist only in two states: the "excited" state (after disintegration of the Michaelis complex) and the relaxed state (after the relaxation of the "energized" state). In the second, "continuous" model, the "excited" enzyme molecule relaxes to equilibrium through a continuous set of intermediate states. Analysis shows that both models lead to the kinetic behavior declining from the required by the Michaelis–Menten equation.

The main feature of the Sidorenko–Descherevsky approach was the statement that conformational relaxation of the functioning enzyme molecule can acquire an important role in enzyme activity. Meanwhile, the mechanism of the chemical transformation of the substrate was considered as the conventional one. The conformational changes in the course of enzyme relaxation influence the activity of an enzyme, but do not take an *immediate* part in performing the elementary chemical act.

The first notion on the deviation of elementary catalytic acts of enzyme reaction, from that prescribed by classical thermodynamic and kinetic approaches, was, probably, formulated in 1971 [19]. It had been shown that the application of basic postulates of activated state theory to the majority of enzyme processes can lead to physically meaningless values of the activation parameters (energy and entropy of activation). It was emphasized that *enzyme functioning is more similar to the work of a mechanical construction than to the catalytic homogeneous chemical reaction.* The selfconsistent phenomenological relaxation theory of enzyme catalysis was proposed in 1972 [20, 21].

What is the principal idea of the relaxation concept? This is not simply a question of the conformational relaxation of the substrate–enzyme complex associated with changing the enzyme catalytic activity. *The substrate binding to an enzyme active center initiates the conformational relaxation acting as the driving force that pushes the chemical system (substrate molecule attached to the catalytic center) along the reaction coordinate.*

Speaking, for certainty, of the "substrate binding" as the factor triggering the conformational transition of a system to the new state of equilibrium, we have to remember other factors. As noted in Section 4.1, any local chemical change in the protein molecule (the substrate or inhibitor binding to the active center, redox change of a group in the prostetic group, ionization of acid or base group, etc.) can lead to the appearance of a conformationally nonequilibrium state. The fast vibrational relaxation ($\tau \cong 10^{-12}$ s) of the active center and its nearest surroundings takes place immediately after the local disturbance, while the structure of the whole protein globule still remains practically the same. However, the structure of the unchanged globule becomes the nonequilibrium one. The new kinetically available equilibrium state for the whole system (the enzyme–substrate complex) will correspond to the conformationally changed structure of the protein globule with the product bound to an enzyme active center. The relaxation concept of enzyme catalysis assumes that *the transformation of a substrate molecule into the product is realized in the course of the enzyme–substrate complex conformational relaxation to the state of equilibrium.* As has been stressed above, this transition can proceed extremely slowly compared with the time scale of vibrational relaxation. In certain cases, this process takes a few hundred milliseconds or even a few seconds (see, for references, [3, 4, 22]).

Carreri and Gratton [23] subdivided the time scale of the internal protein motions into three classes:

(i) the subpicosecond–picosecond;
(ii) the nanosecond–microsecond; and
(iii) the millisecond-second time scales.

Subpicosecond and picosecond motions are related to localized vibrations. According to [23], it appears that the main contribution to absorption in the spectral interval from 1 to 200 cm^{-1} is caused by "the hydrogen bond kinetics of the protein structural elements and of the bound water rather than by the excitation of the protein structure." Although this kind of motion primarily includes the solvent, it probably provides a viscous damping for the fast conformational fluctuations, and thus can play a certain role in the relaxation process. As noted by the authors, the most important time scales are the nanosecond and the microsecond ones. Corresponding motions determine the internal mobility in proteins, as well as in an enzyme action. Otherwise, Carreri and Gratton are sure that "the motions in the millisecond–second time scale are not important for the determination of the catalytic properties of an enzyme." The authors discussed mainly the conformational fluctuations which take place near the conformationally equilibrium state of a protein globule.

According to the relaxation concept [3, 4, 21, 22], a scheme of an enzyme reaction can be written as following

$$E \underset{-S}{\overset{+S}{\rightleftarrows}} ES \rightarrow \tilde{E}P \underset{+P}{\overset{-P}{\rightleftarrows}} \tilde{E} \rightarrow E. \tag{4.2}$$

$$\begin{array}{cccc} a & b & c & d \\ \text{(fast)} & \text{(slow)} & \text{(fast)} & \text{(slow)} \end{array}$$

There are two kind of stages in this extremely simplified scheme:

(i) fast reversible binding/releasing steps (a and c); and
(ii) practically irreversible slow relaxation stages (b and d).

Any relaxation step (e.g., stage b) can be written in the form of a sequence of many reversible elementary acts, each one being the dissociation or formation of a secondary bond, twisting a certain group around a single covalent bond, etc.,

$$ES \; \rightleftarrows \; E'S' \; \rightleftarrows \; E^*S^* \; \rightleftarrows \cdots \rightleftarrows \; \tilde{E}P$$

$$\downarrow\uparrow \qquad \downarrow\uparrow \qquad \downarrow\uparrow \qquad\qquad \downarrow\uparrow \qquad\qquad (4.3)$$

$$E + S \quad E' + S' \quad E^* + S^* \qquad \tilde{E} + P.$$

The chemical transformation $S \rightarrow P$ is just one particular elementary act in the immense set of elementary steps in the scheme (4.3). Of course, in the case of complex reaction, involving the transformation of several components, there may be a number of such elementary acts that represent the catalyzed reaction. However, even in the simplest case of monomolecular reaction the scheme (4.3) is also simplified: the sequence of elementary acts on the path from ES to $\tilde{E}P$ may be branching, scheme (4.4). Hence, there are a set of trajectories leading from ES to $\tilde{E}P$.

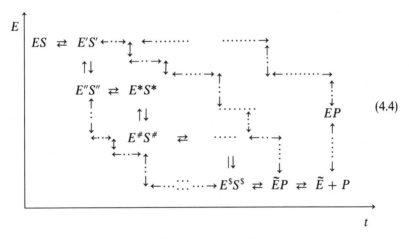

$$(4.4)$$

The possibility of a microscopic description of this relaxation from first principles is practically excluded. Although every elementary act is reversible, the overall process (b), as well as (d), is a practically irreversible one. This circumstance enables us to consider the evolution from ES to $\tilde{E}P$ as a mechanical motion that will stop after reaching an equilibrium state by the enzyme complex. At first glance, the occurrence of the irreversibility in the course of enzyme relaxation is similar to that situation in the classical theory of dynamic systems and physical kinetics. The systems, consisting of a large number of interacting particles and being shifted from the equilibrium, display the irreversible behavior regardless of the reversible nature of each elementary act (such as the elastic collisions of particles, etc.) [24–30].

The evolution of an enzyme complex to the equilibrium state is realized as a random walk through the set of reversible elementary transitions. In general, these elementary steps can also include the reorganization in the immediate low-molecular surrounding of the enzyme molecule. There may be a huge number of trajectories between the initial and final states. We might conclude, therefore, that the existence of kinetically available but different trajectories leading to the equilibrium state will mean that the movement to equilibrium has a purely statistical nature. On the way toward equilibrium, for any elementary step of transformation between two states, $i \to j$, the enzyme complex will overcome a certain potential barrier, ΔU_{ij} (Fig.4.3). According to Kramers' theory of rate processes [31], that considers each elementary act of chemical reaction as the diffusion over a potential barrier, the rate constant for the direct and backward transformations, $i \to j$ and $j \to i$, are

$$k_{ij} = \tau_{ix}^{-1} \exp(-\Delta U_{ij}/RT),$$
$$k_{ji} = \tau_{jx}^{-1} \exp[-(\Delta U_{ij} + \Delta G_{ij})/RT].$$

Here, τ_{ix} and τ_{jx} are the relaxation times of the corresponding structural fluctuations, and ΔU_{ij} and ΔG_{ij} are the energy of barrier and free energy change for the reversible transitions $i \leftrightarrow j$. Bearing in mind the exponential dependence of rate constants on the values of energetic parameters, and taking into account that these parameters differ for the elementary steps lying on different trajectories, we can conclude that the characteristic times of the system relaxation along various trajectories may differ significantly. For this reason, among a huge number of possible trajectories, the only certain ones will dominate in the whole ensemble of possible trajectories corresponding to the enzyme relaxation process. We can thus say that the relaxation process would reveal "deterministic" behavior. So, "mechanistic" analogy seems to be

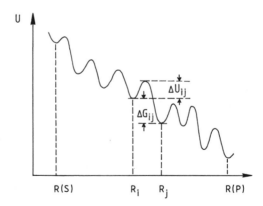

Fig. 4.3. A profile of a system potential energy versus a reaction coordinate corresponding to one of the possible trajectories of the enzyme-catalyzed reaction $S \to P$ (in accordance with [31]).

relevant for an adequate description of the irreversible process of enzyme relaxation associated with the transformation $S \rightarrow P$. Following such a mechanistic approach, we could also say that the "mechanical motion" of a relaxing enzyme molecule is driven by a force arising between the relaxed and unrelaxed regions of a macromolecular complex.

Appealing to the "mechanical" analogy might also seem to be reasonable for another reason. Under certain circumstances the enzymatic reactions of a substrate transformation into a product are reversible processes. As we know, the reversibility is the property of mechanical devices having a limited number of selected (mechanical) degrees of freedom. However, enzymatic paths of the direct ($S \rightarrow P$) and reverse ($P \rightarrow S$), chemical transformations can, in fact, be quite different. There may be several reasons for such a difference. One of them is the existence of the irreversible relaxation step, $\tilde{E} \rightarrow E$, that follows the step of a product dissociation. According to the reaction mechanism (4.3), after releasing a product P ($\tilde{E}P \rightleftarrows \tilde{E} + P$), an enzyme molecule turns out to be in the nonequilibrium state $\tilde{E}$ that will relax to the initial equilibrium state E. If the reverse reaction starts with binding P to an enzyme in its equilibrium form, E, then the "backward" path from P to S will be different as compared with that path which can be simply obtained by the reversion (from right side to left) of processes depicted on the scheme (4.3). Experimental evidence for different pathways for direct and reverse enzymatic reactions will be considered in the next section.

Obviously, not only the paths but also the overall rates of forward and backward processes, $S \xrightarrow{k_{SP}} P$ and $P \xrightarrow{k_{PS}} S$, can be different. Under real experimental conditions, biochemists deal with the ensemble of molecules consisting of a number of enzyme molecules and low-molecular ligands (P, S, etc.). For the whole system (the ensemble of enzyme molecules and their water-soluble ligands), the forward and backward fluxes, J_{SP} and J_{PS}, can be determined as the numbers of "travels" per unit time along the coordinate reaction in both directions, $S \rightarrow P$ and $P \rightarrow S$. The fluxes are determined not only by the apparent rate constants, k_{SP} and k_{PS}, but also by the concentrations of P and S in the solution. The direct and backward fluxes are equal to $J_{SP} = N_S k_{SP}$ and $J_{PS} = N_P k_{PS}$, respectively, where N_S and N_P are the numbers of the enzyme complexes with bound molecules of substrate S and product P. The values N_S and N_P are determined by the concentrations of S and P in the medium and their affinities to the enzyme molecule. Under equilibrium conditions, $N_S k_{SP} = N_P k_{PS}$.

Rigorously speaking, the paths of the direct and reverse processes do not coincide if the whole system undergoing the chemical transformation, $S \rightleftarrows P$, is far away from the thermodynamic equilibrium. In this case only, we can be sure that both processes, forward and backward, will start from the binding S and P to equilibrium form of the enzyme molecule, E. Let us assume, for certainty, that the chemical affinity for the mixture of S and P molecules is positive; these might be realized at the appropriate concentration of S and P in the solution. In this case, characterized by the low "pressure" of P molecules, the probability of initiating the backward step, $\tilde{E} + P \to \tilde{E}P$, will be negligibly small, since the nonequilibrium form of the bare enzyme, $\tilde{E}$, would have time to relax to its equilibrium state, $\tilde{E} \to E$. For this reason, the majority of relatively rare backward events $P \to S$ will start with P binding to the equilibrium form of the enzyme molecule, $E + P \to EP$, and, therefore, the trajectories of both processes, direct and reverse, should be different (scheme (4.4)). Otherwise, if the S–P system is not far from equilibrium, or it is characterized by the negative affinity of the P–S system, being under a rather high "pressure" of P, most of the reverse processes can start before the relaxation of $\tilde{E}$. In this particular case, the backward trajectory may represent, simply, the reversion of the direct sequence (4.3). Another option is that there may be not simply the establishment of dynamic equilibrium between the forward and backward processes, but *stopping* the process in both directions [32].

There were many attempts at a theoretical description of the relaxational motion of the enzyme molecule. In order to describe the behavior of an *individual* enzyme complex, we have to discard the usual approach of the transition-state theory. According to this theory [33], an elementary act of chemical transformation proceeds via a so-called activated complex which is in a state of thermodynamic quasi-equilibrium. The classical chemical kinetics always consider the equilibrium state as a dynamic process. At equilibrium, the rates of direct and reverse reactions are equal but both processes never stop. In a mechanical process, equilibrium means the end of any movement.

For a description of enzyme catalysis, it was very tempting to apply the above-mentioned theory of rate processes developed by Kramers as early as 1940 [31]. Recently, this approach has been reviewed in depth and further advanced by Gavish [34]. According to Kramers' theory, the $S \to P$ transformation can be simulated by the diffusion of a Brownian particle in the potential field. The force, driving the process, arises from random fluctuations in the structure of a molecular system. The motion of a "particle" (substrate) is facilitated by a fluctuating random force, and the kinetic energy of this "energized particle" dissipates due to the action of a friction force. The protein moiety is thus considered as a pressure-fluctuating environment. It has been shown in the work cited above [34] that for a volume-dependent reaction the pressure fluctuations can enhance the rate constants of an enzyme reaction in both directions.

The immediate relation to the relaxation concept of enzyme catalysis [3, 4, 20, 21] is the theory of rate processes that has been developed by Fain [35]. In this paper, Fain states that the conventional approach to the kinetics of chemical reactions does not hold true for highly ordered macromolecular structures. The traditional approach to the problem implies that all vibrational modes undergo fast relaxation to thermal equilibrium. In the conventional approach to chemical kinetics, the changes in electronic states and nuclear vibration amplitudes were considered separately. Fain has proposed the self-consistent description of simultaneous changes taking place in the electron and nuclear subsystems, and has obtained equations describing the peculiar kinetic behavior of enzymatic processes. He concluded that *for highly ordered macromolecular systems, in the course of a radiationless relaxation the surplus of electron energy may be transformed into the energy of a coherent vibrational motion without its fast dissipation into heat.* Following Fain, the relaxational transition "*b*" in our scheme (4.2) could be initiated by the excitation of coherent vibrations as a result of substrate binding. The probability of their excitation by thermal degrees of freedom is negligibly low.

Using the language of classical chemical kinetics, the relaxation concept of enzymatic reactions can be roughly expressed in the following way. A substrate–product transformation requires the overcoming of a certain activation barrier. The excitation of coherent nuclear vibrations is equivalent to the excitation of a specific degree of freedom coinciding with the reaction coordinate. This is similar to the creation of the inverse populations of higher and lower levels in the system. In other words, this specific degree of freedom, that accumulates energy in the form of coherent vibrations, can be characterized by very high local temperature. The probability of its "changing for" noncoherent vibrations is negligible. The movement along the most probable relaxation path (determined by the system construction) leads to the substrate–product transformation, which is realized at high "local temperature." This facilitates the overcoming of the reaction activation barrier that cannot be regarded in this case as an obstacle.

4.4. Protein Dynamics and Enzyme Functioning

As we stated in the previous section, in the foundation of the relaxation concept of enzymatic catalysis lies a postulate of the existence of rather long-living nonequilibrium states of individual macromolecules that appear in the course of an enzyme catalyzed chemical reaction. According to this concept, the conformational relaxation of a protein globule with an attached substrate molecule is itself an elementary step of the enzymatic reaction. Conformational changes, initiated by the substrate binding, engage not only a protein globule, but also inevitably include the chemical changes necessary for the substrate–product transformation. Thus, the rate of the substrate–product transformation should be determined by the protein globule relaxation rate. A good deal of experimental evidence from enzymology suggests, in particular, that in many cases, conformational changes causing the product release are the rate-limiting steps of the product formation [36, 37]. The reader can find biophysical evidence of the existence of long-living nonequilibrium states in original and reviewing articles (see, for references, [38, 39]). To illustrate our line of argument in support of the relaxation concept, we will consider below some of these experimental data. We want, however, to start with certain theoretical questions concerning the role of fluctuations in protein structures and other relevant dynamic factors that might play an important role in enzyme functioning.

4.4.1. Theoretical Aspects of Protein Structural Dynamics

In Chapter 2, considering the conventional mechanism of the indirect coupling of chemical reactions, we pointed out that the energy for the elementary steps of chemical transformations associated with overcoming potential barriers was provided by the thermostat (environment) supplying appropriate "hot" particles that occur as a result of thermal fluctuations. As concerns the problem of enzymatic catalysis, the following question arises. What is the role of structural fluctuations in providing the molecular mechanism by which the enzyme molecule encourages a substrate to chemical transformation: either the enzyme molecule prompts the picking out of rare but large fluctuations, thus helping to overcome the potential barrier, or fluctuations in the protein structure lower the barrier?

Phenomenological models to overcome the energy barrier of the enzymatic reaction by using structural fluctuations were suggested by many authors (for references, see [40, 41]). One of them was the model proposed by Gavich and Werber [42]. Being based on Kramer's theory of the rate of chemical processes, this approach suggested that the dynamic behavior of the enzyme–substrate complex is governed by a protein globule interaction with the solvent molecules through the action of two competing processes: random collisions and frictional forces. The former factor leads to increasing the kinetic energy of the various modes (or, say, quasi-particles) in the protein structure, and the action of the latter causes the dissipation of energy by viscous damping. According to the mechanism suggested in [42], the energy–substrate complex can overcome the potential barrier as a result of structural fluctuations depending on the solvent viscosity. Some other aspects of this and similar widely spread viewpoints on the protein, as the matrix serving the role of a fluctuating "reservoir" that generates catalytic configuration, were reviewed by several authors in [41].

Promising perspectives for disclosing the secrets of protein dynamics has brought in the use of theoretical methods based on direct calculations. More than 15 years ago, with the development of powerful computer techniques, there appeared the possibility of applying molecular dynamics approaches to the study of globular protein internal dynamics [43–50]. These methods permit the calculation of atomic trajectories for all protein atoms from first principles, by solving the classical equations of motion using the empirical interatomic potential energy function that determines the forces acting on all atoms.

Successful applications of the molecular dynamics approach, to the description of several folded globular proteins with known atomic coordinates, revealed rather interesting features of the protein internal mobility. The direct simulation of protein behavior suggested that the protein interior in many aspects is fluid-like, and the local motions of atoms have a diffusional character. Although the average positions of atoms are confined to their neighborhood, the fluctuations in atom positions reveal a diffusional character: the dynamics of their displacements is determined by collisions with neighboring atoms [44]. The theoretical study of atomic fluctuations in a bovine pancreatic tripsin inhibitor has revealed a wide range of relaxation times ($\tau \simeq 0.2$–10 ps) corresponding to the correlation functions of atomic displacements [46]. This result suggests the superposition of two types of internal motions:

(i) high-frequency oscillations (subpicosecond) of the individual atoms; and
(ii) lower frequency "collective" modes (1–10 ps) corresponding to the motion of rigid "effective particles" consisting of a group of atoms.

In the second case the energy of vibrational modes can be localized for relatively long time periods in proteins.

One of the most interesting and promising theoretical investigations, explaining the role of structural fluctuations in catalyzing chemical reactions, was direct molecular dynamics simulation of the transport step in the catalytic reaction of lysosyme performed by Warshel [49]. He calculated the dynamics of proton transfer from glutamate-35 of lysosyme to the oxygen atom of a substrate bound to the enzyme active center. The very approach used in this work takes into account the fluctuations in the positions of all atoms and solvent molecules surrounding the protein globule, thus giving the "real-time" picture of the energy profile along the reaction coordinate and characterizing the dynamics of chemical events in the active center. As we know, this work was probably the first computer simulation of the dynamics of actual enzyme catalyzed bond-breaking events, that clarified the type of fluctuations involved in enzymatic reactions. Warshel has found that a key dynamic factor in the reaction of proton transfer was the enzyme electrostatic potential. The rate constant was determined by two main factors stabilizing the transient state charge distribution: the pure contribution of the reacting bonds and that of the protein surroundings. Simulation and direct numerical evaluation of the probability of reaching the transition state demonstrated that "the barrier for proton transfer, $\Delta E^{\ddagger}$, fluctuates as a function of the structural fluctuations of the protein globule. The barrier becomes small only when a fluctuation of a protein generates 'catalytic configuration' that stabilizes the transition state" [49].

These conclusions follow directly from the simulation of a time-dependent proton energy profile along the reaction coordinate. Figure 4.4 demonstrates how random thermal fluctuations cause a temporary lowering of the potential barrier on the line between two extreme positions for a proton. Due to structural fluctuation, at a certain moment of time ($t \cong 16.5$ ps) there appears the transient configuration for which an oxonium state becomes more stable

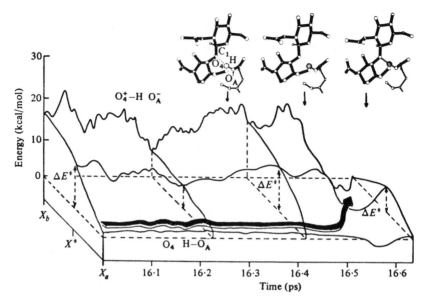

Fig. 4.4. The time-course simulation of potential energy changes for proton transfer along a trajectory from $Glu_{35}(O_A)$ to the $O_{(4)}$ position of the glucosidic bond in the catalytic reaction of lysozyme. (From Warshel and Russel. *Quart. Rev. Biophys.* Vol. 17, pp. 283–427. Reproduced with permission). The reaction coordinate $X = X_a$ corresponds to the main contribution of the resonance structure O_4H-O_A, while coordinate $X = X_b$ corresponds to the ionic resonance structure $O_4^+-HO_A^-$. The points of minima in the value of the activation barrier $\triangle E^{\ddagger}$ correspond to increasing the probability of the proton transfer.

than a carboxylic acid. In this configuration the energy of the transient state, $E^{\ddagger}$, decreases; thus the fluctuation increases the probability of passing the transient state and reaching a new position for a proton. The gradient of proton potential energy along the reaction coordinate, $\partial E/\partial X$, represents a force that pushes the proton to a new position.

Proton transport itself occurs very fast (a few femtoseconds); its velocity was evaluated about 10^{13} Å/s (or about 10^3 m/s, which is a value of the same order of magnitude as the sound velocity in liquids). The rate-limiting step of the overall process is determined by the waiting time for this relatively rare event. Reaching an optimal configuration in the course of the concerted movement of a large number of atoms may take a rather long time. Concerning the role of fluctuations in enzymatic catalysis, Warshel suggested that "the productive bond-break fluctuations of the protein occur only after a fluctuation of the protein reaches the neighbourhood of a catalytic configuration and not as a completely concerted motion of the reactive bonds and the protein configuration." According to [49], the change in the reacting bond occurs in 0.1 ps, while a much longer time (about 4 ps) is required for the structural rearrangement of the protein. We can thus say that *the elementary step of the catalytic act occurs much faster than the following step of the enzyme globule relaxation to its ground state.*

It should be noted, however, that the above-considered example of the direct theoretical simulation of catalytic events relates to the description of one certain stage of the catalytic process with a relatively low potential barrier. According to [49, 50], theoretical investigation of the processes with high activation barriers was possible by separately evaluating the probability of reaching the transient state and the dynamic behavior of the transient complex. For proton transfer reaction, it has been found that fluctuations of the protein globule may contribute to catalysis due to their influence on the probability of reaching the transient state.

Like most other examples from molecular dynamics simulation (see, for references, [44–48]), this theoretical study deals with the fluctuations and rather small deviations from the protein equilibrium state. From the viewpoint of the relaxation concept of enzyme catalysis, the most interesting case should represent the dynamic processes in proteins far removed from equilibrium as a result of fast local perturbations (substrate or coenzyme binding to the active center, etc). The relaxation of the protein globule to the new state of equilibrium should involve not only the fluctuations of atoms (or certain atomic groups) around their quasi-equilibrium positions, but could also lead to rather significant changes in the atom coordinates in the whole protein globule. As we have noted above, there are several reasons to believe that this relaxation would take much more time than relatively fast ($\tau \simeq$ 0.2–10 ps) thermal fluctuations in the interior of a protein globule. One of them might be the excitation of slowly relaxing degrees of freedom like those "collective" quasi-particles that have been theoretically predicted for highly ordered molecular systems by Davydov [51] and Fain [35].

Another possible reason for the slow relaxation rate may be of a purely statistical nature. As was noted in the preceding section, the relaxation of the enzyme complex to a new equilibrium state can be regarded as traveling along the trajectory consisting of the elementary reversible steps of the structural or chemical transformation. The reversibility of the elementary steps implies that there is a local thermal equilibrium at any intermediate state, while the apparent "deterministic" behavior of the system with the relatively long-living states is explained by kinetic factors. In order to pass the trajectory, $ES \rightarrow EP$, that represents the "chain" consisting of the immense set of equilibrium intermediate states, the system will make an enormous number of futile attempts before reaching its appropriate final equilibrium state. This may be the reason for a dramatic increase in the relaxation time of the overall process. This is the case, for example, for the transition between the regular double helix and a cruciform structure in superhelical DNA. Theoretical calculations based on a purely statistical approach, that were performed by Vologodskii and Frank-Kamenetskii in [52], are in a good agreement with the experimental observations that the relaxation time of this process may reach many hours, or more [53, 54].

4.4.2. Experimental Evidence for Protein Nonequilibrium States and Their Evolution in the Course of Enzyme Turnover

Let us now consider experimental evidence of the existing nonequilibrium states in proteins that may be relevant to enzyme functioning. Certain predictions of the relaxation concept, such as the conformational changes of a protein globule in the course of the enzyme cycle, and the difference between the forward and backward reaction pathways, have been verified in [55]. In this investigation, using a stop-flow technique for initiating the reaction, the authors monitored parameters reflecting the structural properties of the protein globule and enzyme catalytic activity in the course of forward and backward reactions catalyzed by soluble cytoplasmic malatedehydrogenase during a single turnover of enzyme cycle

$$HOOC-CHOH-CH_2-COOH + NAD^+$$

$$L\text{-malate}$$

$$\rightleftarrows HOOC-CO-CH_2-COOH + NADH + H^+.$$

$$\text{oxaloacetate}$$

This particular reaction has been chosen for the reason of its high value of standard chemical affinity for this reaction ($\mathscr{A}^0 = -7.1$ kcal/mole). As we noted above, due to this circumstance the system behavior can reveal the deviation from that prescribed by the classical Arrhenius mechanism. The conformational changes in malatedehydrogenase were tested by measuring the average life-time of intrinsic tryptophane fluorescence ($\bar{\tau}_f$). This parameter is known to be sensitive to the immediate surrounding of tryptophane residues. The chemical transformation of the substrate was detected from changes in the coenzyme redox state measured in terms of the sample optical density at 340 nm (NADH absorption maximum).

The turnover numbers for the forward and backward reactions, measured under the surplus of the corresponding substrate and coenzyme, were found to be $5.5 \cdot 10^3$ min^{-1} and $2.3 \cdot 10^4$ min^{-1}, respectively. This means that one act of the forward reaction (L-malate $\rightarrow$ oxaloacetate) takes 10.9 ms, and that of the backward reaction (oxaloacetate $\rightarrow$ L-malate) takes 2.6 ms.

Figure 4.5(A) demonstrates the time courses for the fluorescence parameter $\bar{\tau}_f$ and NAD$^+$ reduction kinetics measured at 20 °C after fast initiation (the mixing dead time was about 1.2–1.5 ms) of the forward reaction. The general features of the kinetic picture did not depend on the method of the reaction initiation: by the addition of substrate to the mixture of enzyme and coenzyme (curve 1), or by the addition of the coenzyme and substrate mixture to the enzyme solution (curve 2). From Figure 4.5(A) it follows that the structurally-dependent parameter $\bar{\tau}_f$ decreases, reaching its minimal value, $\bar{\tau}_f \cong 2.4$ μs, at the moment of time $t_R \cong 6$ ms after initiating the reaction. At the end of the first enzyme cycle ($t_R \cong 10$–11 ms) the parameter $\bar{\tau}_f$ returns to its initial value, $\bar{\tau}_f \cong 3.05$ μs, characteristic for the enzyme–NAD$^+$ complex,

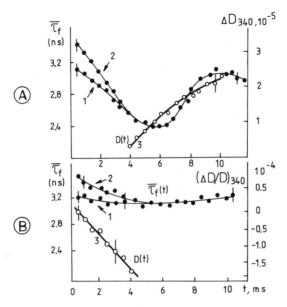

Fig. 4.5. The time-courses for tryptophane fluorescence parameters $\bar{\tau}_f$ and NAD$^+$ reduction (A) or NADH oxidation (B) reactions catalyzed by malatedehydrogenase (after [55]).

and then decreases again with a new turn of the enzyme cycle. The reduction of NAD$^+$ begins only 3–4 ms after the reaction initiation (Fig. 4.5(A), curve 3). Thus, the product formation occurs only after a rather significant structural change in the enzyme globule that reveal itsself in the time course of the intrinsic fluorescence parameter $\bar{\tau}_f$.

A quite different kinetic behavior is displayed by the backward reaction (oxaloacetate $\rightarrow$ L-malate, $\mathscr{A}^0 = 7.1$ kcal/mole). The course of this reaction is practically not accompanied by a noticeable decrease in the $\bar{\tau}_f$ value; the NADH oxidation begins immediately after mixing and proceeds during several enzyme cycles (Fig. 4.5(B)).

All these data can be considered as direct evidence of the fact that the soluble cytoplasmic malatedehydrogenase catalyzes the forward and backward reactions being in different conformational states. The pathways for these processes do not coincide.

The data obtained in [55] can also elucidate the mechanism of "hysteresis" or "mnemonic" enzyme activity. Generally, there are two options for explaining the activation of such enzymes:

(i) the gradual increase in the activity of each individual enzyme molecule with each new turnover; and

(ii) the increase in the number of active enzyme molecules in the initially inactive population of enzyme molecules.

It follows from the data presented in [55] that in the first cycle of enzyme turnover, only an insignificant portion of the malatedehydrogenase molecules take part ($\approx 0.1\%$), but all of these enzyme molecules reveal their maximal activity from the very beginning. This is the typical behavior of hysteresis enzymes which can be in two discrete states—active and inactive. In the course of the reaction the number of active molecules increases. Each active enzyme molecule, taking part in the act of the substrate–product transformation, is removed from the population of potentially active but non reacted molecules. Therefore, the equilibrium between two populations becomes shifted, and the number of active enzyme molecules increases with each enzyme turnover until all the enzyme molecules become active.

The rate of enzyme turnover is usually determined, not simply by the duration of the substrate–product transformation at the enzyme active center, but by the rate of protein globule relaxation. There is much experimental evidence for this statement. For example, a kinetic and structural study of cytochrome C modified by the ruthenium label specifically at the imidazole moiety of histidine 33, demonstrated that the rate of intramolecular electron transfer, $Ru(2) \rightarrow Haem(3)$, $k \cong 55$ s^{-1} [56], is on the same time scale as the rate of conformational changes occurring within the cytochrome C molecule [57, 58]. Together with other experimental results (see, for references [38, 39]), this is an obvious indication of the protein conformational changes as the rate-limiting step in protein functioning.

Finally, we want to consider a simple but very explicit example illustrating the crucial role of structural reorganization in the functioning of even very simple molecular systems. Following the data and arguments presented in the comprehensive review by Gutman and Nachliel [59], we will consider the mechanism of proton propagation in aqueous solutions. At room temperature each water molecule forms two to three hydrogen bonds with its immediate neighbors. The average distance between the oxygen atoms of neighboring water molecules is about 2.75 Å. This provides the establishment of a flexible (dynamic) network of hydrogen-bonded water molecules spreading over the whole volume of the aqueous solution. The network can incorporate the hydrogen-bonded proton hydrates and some other structures. The lifetime of the hydrogen bond is rather short, $\tau_{HB} \cong 2.3$ ps. However, this time is quite enough to perform the elementary step of proton transfer between two hydrogen-bonded water molecules which takes about 45 fs. This provides a specific mechanism of protons propagation in water which differs from proton self-diffusion.

According to [59], at 25 °C the diffusion coefficient of protons in water is $D_{H^+} \cong 9.28 * 10^{-5}$ cm^2/s, while the self-diffusion of water molecules is three times lower, $D_{H_2O} \cong 2.8 * 10^{-5}$ cm^2/s. The hydrogen-bonded network, that restrains the self-diffusion of water molecules, paves the way for the rather fast proton transfer along the hydrogen bonds. Curiously, ordering water molecules within the network in itself does not accelerate proton transfer. Actually, according to recent data [60], the rate of proton diffusion in the rigid lattice of ice is smaller than that in supercooled water, $D_{H^+(ice)} = 0.7 * 10^{-5}$ cm^2/s and $D_{H^+(aq)} = 4.1 * 10^{-5}$ cm^2/s (at -10 °C). This means that it is the structural mobility (reorientation of water molecules), but not fixing the network structure, that provides the conditions for the propagation of protons. We can consider this example as an explicit illustration of the fact that the very possibility of structural reorganization does not hinder but only hastens a chemical process (proton transfer).

Let us consider in more detail the reaction of proton transfer from an acid AH to a base B, $A—H + :B \rightleftarrows A: + H—B$. This process can be subdivided into three main stages:

(i) a diffusion-controlled step of hydrogen-bond formation, $A—H + :B \rightleftarrows A—H\cdots B$;

(ii) a proton transfer accompanied by the formation of a new complex, $A—H\cdots B; \rightleftarrows A\cdots H—B$; and

(iii) a diffusion controlled step of complex dissociation, $A\cdots H—B \rightleftarrows A: + H—B$.

Immediately after a proton jump, $A—H + :B \rightarrow A: + H—B$, a proton transferred still "remembers" its previous position, since traveling along the hydrogen bond it leaves behind the unpaired electronic orbitals. After a proton transfer from a donor H_3^+O molecule to an acceptor H_2O molecule, about 36% of the full protonic charge transferred is neutralized by the unpaired electronic orbitals of acid, $A:$, oriented toward an acceptor molecule, HB. Having a "memory" of ionic defect left on a donor, a proton can return to its original position, $A: + H—B \rightarrow A—H + :B$. This is the case, for instance, in the rigid ice-like structures characterized by the hindered rotation of reacting molecules. The randomization of a direction of unpaired orbitals due to the fast rotation of water molecules in liquids will preclude the process of backward proton transfer, thus prompting their traveling along the "flexible" hydrogen-bonded network. Therefore, the experimentally observable overall time of proton transfer from AH to B should be determined by the correlation time of the molecule rotational motion that determines the life-time of hydrogen bonds. The faster molecules rotate, the more rapidly occurs the propagation of protons. As Gutman and Nachliel noted [59], "It is the impermanence of a structure in water which makes the proton so diffusible." The example above illustrates two points that, as we will see below, acquire a special importance for explaining the functioning "molecular" devices:

(i) *the structural flexibility and mobility* at the molecular level (e.g., random tumbling of water molecules) *facilitates the functioning* of the molecular systems (e.g., proton transfer); and

(ii) *the overall turnover times of the "molecular" devices are determined by the time of structural relaxation* (e.g., the correlation time of the H_2O molecules rotation, $\tau \approx 2$–3 ps, but not the duration of a proton-jumping elementary step, $\tau \approx 45$ fs, is the factor that determines the overall rate of proton propagation).

References

1. E.M. Kosower (1962), *Molecular Biochemistry*, McGraw-Hill, New York.
2. A. Fersht (1985), *Enzyme Structure and Mechanism*, 2nd ed., Freeman, New York.
3. L.A. Blumenfeld (1981), *Problems of Biological Physics*, Springer-Verlag, Heidelberg.
4. L.A. Blumenfeld (1983), *Physics of Bioenergetic Pocesses*, Springer-Verlag, Heidelberg.
5. J. Ricard, J. Buc, and J.C. Mennier (1977), *European J. Biochem.* **80**, 581–592.
6. J. Buc, J. Ricard, and J.C. Mennier (1977), *European J. Biochem.* **80**, 593–601.
7. D.E. Koshland, Jr. (1962), *J. Theoret. Biol.* **2**, 75.
8. D.E. Koshland, Jr and R.E. Neet (1968), *Ann. Rev. Biochem.* **37**, 672.
9. S. Milstein and L.A. Cohen (1970), *Proc. Nat. Acad. Sci. USA* **67**, 1143–1147.
10. R. Lumry (1959), *The Enzymes*, vol. 1, Academic Press, New York, p. 157.
11. E. Bauer (1935), *Theoretical Biology*, VJEM Press, Moscow.
12. N.I. Kobosev (1960), *Zh. Fiz. Chim.* **34**, 1443.
13. Yu.V. Medvedev (1930), *New Ideas in the Enzyme Science* (in Russian), USSR Academy of Sciences, Moscow.
14. Yu.J. Churgin, D.S. Chernavsky, and S.E. Shnoll (1967), *Mol. Biol.* (*USSR*) **1**, 419–424.
15. S.E. Shnoll (1967), In: *Oscillatory Phenomena in Biological and Chemical Systems*, Nauka, Moscow, pp. 22–41.
16. E.A. Moelwyn-Hughes (1959), In: *The Enzymes*, Vol. 1, Academic Press, New York, p. 28.
17. C.N. Hinshelwood (1926), *Proc. Roy. Soc.* **A113**, p. 230.
18. N.P. Sidorenko and V.I. Descherevsky (1970), *Biofizika* (*USSR*) **15**, 785–792.
19. L.A. Blumenfeld (1971), *Biophysics* (*USSR*) **16**, 724–727.
20. L.A. Blumenfeld (1972), *Biophysics* (*USSR*) **17**, 954–959.
21. L.A. Blumenfeld (1976), *J. Theoret. Biol.* **58**, 269–284.
22. E.R. Henry, J. Hofreichter, and W. Eaton (1987), In: *Structure, Dynamics and Function of Biopolymers*, Springer Series in Biophysics, vol. 1, Springer-Verlag, New York, pp. 20–24.
23. G. Carreri and E. Gratton (1986), In: *The Fluctuating Enzyme* (G.R. Welch, Ed.), Wiley, New York, pp. 227–262.
24. A.N. Kolmogorov (1954), *Proc. Acad. Sci. USSR* **98**, 527–531.
25. V.I. Arnold and A. Avez (1968), *Ergodic Problems of Classical Mechanics*, Benjamin, New York.
26. S.P. de Groot and P. Mazur (1962), Non-equilibrium Thermodynamics, North-Holland, Amsterdam.
27. R. Balescu (1975), *Equilibrium and Non-equilibrium Statistical Mechanics*, Wiley–Interscience, New York.
28. H. Haken (1978), *Sinergetics*, Spinger-Verlag, Berlin.
29. I. Prigogine (1980), *From Being to Becoming: Time and Complexity in Physical Sciences*, Freeman, San Francisco.
30. F.C. Moon *Chaotic Vibrations*, Wiley–Interscience, New York.
31. H.A. Kramers (1940), *Physica*, **7** (1940), 305.
32. L.V. Belovolova, L.A. Blumenfeld, D.Sh. Burbaev, and A.F. Vanin (1975) *Molec. Biol.* (*USSR*) **9**, 934–940.

33. S. Glasstone, K.J. Laidler and H. Eyring (1941), *The Theory of Rate Processes*, McGraw-Hill, New York.

34. B. Gavish (1986), In: *The Fluctuating Enzyme* (G.R. Welch, Ed.) Wiley, New York, pp. 262–339.

35. V.M. Fain (1976), *J. Chem. Phys.* **65**, 1854–1866.

36. P.D. Boyer (1965), In: *Dynamics of Energy-Transducing Membranes* (L. Ernster, R.W. Estabrook and E.C. Slater, Eds.) Elsevier, Amsterdam, p. 389.

37. W.W. Cleland (1975), *Acc. Chem. Res.* **8**, 145.

38. L.A. Blumenfeld and R.M. Davydov (1979), *Biochim. Biophys. Acta* **549**, 225–240.

39. L.A. Blumenfeld, D.S. Burbaev, and R.M. Davydov (1986), In: *The Fluctuating Enzyme*, (G.R. Welch, Ed.) Wiley, New York, pp. 369–401.

40. B. Somogyi, G.R. Welch and S. Damjanovich (1984), *Biochim. Biophys. Acta* **768**, 81–112.

41. *The Fluctuating Enzyme* (1986), (G.R. Welch, Ed.) Wiley, New York.

42. B. Gavish and M.M. Werber (1979), *Biochemistry* **18**, 1270–1975.

43. *Models for Protein Dynamics* (1976), (H.J.C. Berendsen, Ed.) CECAM, University of Paris IX, France.

44. J.A. McCammon, B.R. Gelin, and M. Karplus (1977), *Nature (London)* **267**, 585–590.

45. M. Karplus and J.A. McCammon (1981), *CRC Crit. Rev. Biochem.* **9**, 293–349.

46. S. Swaminathan, T. Ichiye, W. van Gunsteren, and M. Karplus (1982), *Biochemistry* **21**, 5230–5240.

47. R.M. Levy, D. Perahia, and M. Karplus (1982), *Proc. Nat. Acad. Sci. USA* **79**, 1346–1350.

48. M. Levitt and R. Sharon (1988), *Proc. Nat. Acad. Sci. USA* **85**, 7557–7561.

49. A. Warshel (1984), *Proc. Nat. Acad. Sci. USA* **81**, 444–448.

50. A. Warshel and S.T. Russel (1984), *Quart. Rev. Biophys.* **17**, 283–427.

51. A.S. Davydov (1982), *Biology and Quantum Mechanics*, Pergamon Press, Oxford.

52. A.V. Vologodskii and M.D. Frank-Kamenetskii (1983), *FEBS Lett.* **160**, 173–176.

53. K. Mizuucki, M. Mizuucki and M. Gellert (1982), *J. Mol. Biol.* **156**, 229–243.

54. V.I. Lyamichev, I.G. Panyutin, and M.D. Frank-Kamenetzkii (1983), *FEBS Lett.* **153**, 298–302.

55. L.A. Blumenfeld and P.G. Pleshanov (1987), In: *Structure, Dynamics and Function of Biomolecules* (A. Ehrenberg, A. Graslund, and L. Nilsson, Eds.) Springer-Verlag, Berlin, pp. 171–175.

56. S.S. Isied, C. Kuehn, and C. Worosila (1984), *J. Amer. Chem. Soc.* **106**, 1722–1726.

57. C. Creutz and N. Suten (1973), *Proc. Nat. Acad. Sci. USA* **70**, 1701–1703.

58. C. Creutz and N. Suten (1973), *J. Biol. Chem.* **249**, 6788–6795.

59. M. Gutman and E. Nachliel (1990), *Biochim. Biophys. Acta* **1015**, 391–414.

60. E. Pines and D. Huppet (1985), *Chem. Phys. Lett.* **116**, 295–300.

61. A.A. Timoshin, A.N. Tikhonov, and L.A. Blumenfeld (1984), *Biophysics (USSR)* **29**, 338–340.

CHAPTER 5

Energy Transduction in Biological Membranes

5.1. Introduction: Two Views on the Problem of Energy Coupling in Biomembranes

Oxidative phosphorylation and photophosphorylation are the major processes providing the ATP formation in the energy-transducing intracellular organelles (mitochondria, chloroplasts, and chromatophopres) of living cells. After the pioneering works of Engelgardt, Szent-Gyorgyi, Warburg, Belitser and Tsybakova, and Lipmann, who clarified the role played by the ATP molecule as the universal energy currency in the majority of biochemical processes, a multitude of experimental and theoretical studies in the field of bioenergetics were carried out during the last 50 years (see reviews on the history of bioenergetics in [1–14]). Despite the great volume of publications, the molecular mechanism of the most important bioenergetic process, i.e., the membrane phosphorylation (ATP synthesis from ADP and P_i), still remains a point of lively debate in the scientific literature. As a matter of fact, the mechanism of ATP synthesis is essentially a problem of coupling between the energy-donating (electron transport) and energy-accepting (ADP + $P_i \rightarrow$ ATP + H_2O) processes.

Figure 5.1 depicts the sketches of the cross sections of chloroplasts and mitochondria, i.e., the energy-transducing organelles of plant and animal cells. The great mass of experimental data obtained in several laboratories has given detailed information on the composition and molecular structure of electron transport chains (ETC) in chloroplasts, mitochondria, and bacteria, and localization of the so-called energy-coupling sites in ETC (see, for references, [11–16]). Figures 5.2–5.5 demonstrate the schemes of electron transport chains in chloroplasts and mitochondria. Some electron transport processes are coupled with the protonation/deprotonation of certain electron carriers. Due to the asymmetrical arrangement of electron transport chain components in coupling membranes, the functioning of chloroplasts and mitochondrial ETC causes the transmembrane translocation of protons across the thylakoid membranes and inner mitochondrial membranes.

ATP synthesis in biomembranes is not directly caused by electron trans-

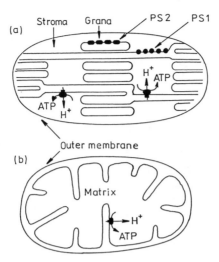

Fig. 5.1. Schematic representation of (a) chloroplast and (b) mitochondrion cutaways.

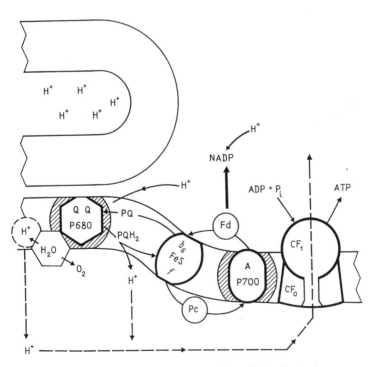

Fig. 5.2. A scheme of the possible arrangement of the chloroplast's electron transport and ATPsynthase complexes in the thylakoid membrane (after [206, 207]).

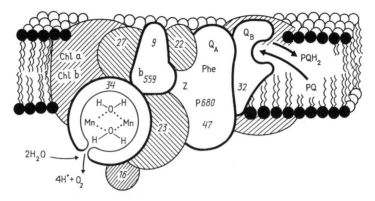

Fig. 5.3. A scheme of the photosystem 2 complex arrangement in the thylakoid membrane. P680 is the reaction center; electron acceptors: Phe is the pheophytin; Q_A and Q_B are bound plastoquinone; PQH_2 and PQ are free plastoquinol (reduced form) and plastoquinone (oxidized form) molecules, respectively; b_{559} is cytochrome b_{559}; and Z is the electron donor for the oxidized reaction center $P680^+$ (after [206, 207]).

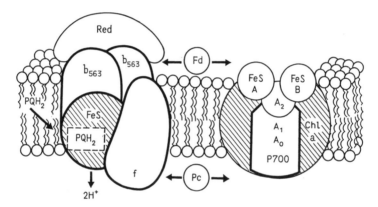

Fig. 5.4. A scheme of the photosystem 1 and the b_6/f complexes arrangement in the thylakoid membrane. Photosystem 1: P700 is the reaction center; A_0, A_1, and A_2 are electron acceptors; and FeS_A and FeS_B are two forms of bound ferredoxin. b_6/f complex: f is the cytochrome f; b_{563} is cytochrome b_{563}; FeS is the Rieske electron transport protein; PQH_2 is the plastoquinol molecule bound to Rieske center; Red is ferredoxin-NADP-reductase; Fd is water soluble ferredoxin; and Pc is plastocyanin.

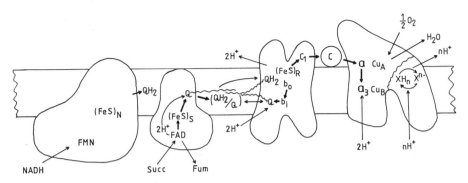

Fig. 5.5. Scheme of a possible arrangement of four electron transport complexes in the cristae membrane of mitochondria. $(FeS)_N$, $(FeS)_S$, and $(FeS)_R$ symbolize the nonhaem Fe–S centers of the Green complexes 1, 2, and 3, respectively; FMN is flavinmononucleotide; FAD is flavinadenindinucleotide; c_1 is cytochrome c_1; c is cytochrome c; a and a_3 are the cytochrome a and a_3 of the cytochrome oxidase complex.

port reactions. Energy-donating and energy-accepting processes are realized as time-separated and spatial-separated reactions. All the data available now demonstrate that ADP is the immediate acceptor that covalently binds P_i, without forming any phosphorylated intermediate. However, the redox transformations of ETC carriers in energy-transducing membranes leads indirectly to ATP synthesis. For many years, it was known that energy released in the course of electron transfer from a low-potential electron carrier to a high-potential component of ETC can be stored in the intermediate form even in the absence of the phosphorylation substrates (ADP and P_i). Electron transport and ATP formation can be uncoupled. In uncoupled energy-transducing organelles electron transport occurs without ATP synthesis. Energy released in the course of oxidative processes dissipates into heat. Energy-transducing membranes lose their coupling ability more easily than the electron transport activity. For instance, mitochondria passes into the uncoupled state even as a result of relatively weak damage: heating up to 50 °C, a repeated freezing and thawing procedure, aging. There is a large group of compounds (called "uncouplers") whose addition to energy-transducing organelles suppresses the ATP formation without the inhibition of electron transport chains.

On the other hand, it is possible to create conditions which enable us to catalyze the ATP formation without electron transport. It was demonstrated in the early 1960s that the illumination of chloroplasts in the absence of phosphorylation substrates (ADP and P_i) led to the formation of a certain energized compound (or state), which was capable of providing ATP synthesis in the dark after ADP and P_i addition [17, 18]. This post-illumination ATP synthesis is one of the most explicit demonstrations of the existence in chloroplasts of a high-energy state formed in the light. Later, Jagendorf and

Uribe demonstrated that such (or similar) a macroergic state can be created artificially as a result of the acid-base transition: chloroplasts preincubated at a low pH value in the dark were able to produce ATP from ADP and P_i after a fast pH increase [19]. The ATP formation, coupled to artificially generated macroergic states in chloroplasts, mitochondria, and submitochondrial particles, chromatophores, and reconstituted model systems (lipid vesicles encrusted with ATPsynthase complexes but free of ETC) was studied in more detail by several groups of investigators (see, for references, [20–35]).

All these data unambiguously indicate the existence of nonphosphorylated high-energy intermediate(s) coupling the energy-donating reactions of electron transport with ATP formation. This "energized" intermediate is also called the "primary macroerg," $\tilde{X}$. The formal general scheme of membrane phosphorylation can be written as follows:

$$
\begin{array}{lll}
D_{red} + A_{ox} \rightarrow & \rightarrow \tilde{X} \rightarrow & \rightarrow ATP + H_2O \\
D_{ox} + A_{red} \leftarrow & \leftarrow X \leftarrow & \leftarrow ADP + P_i
\end{array}
\qquad (5.1)
$$

All hypotheses concerning the mechanisms of membrane phosphorylation differ on the assumptions concerning the nature of the primary macroerg $\tilde{X}$. The coupling mechanism based on the idea that this intermediate may represent a chemical compound (e.g., a specific ligand bound to an electron carrier) was first clearly formulated by Slater in 1953 [36] and elaborated on in other laboratories [37]. However, all efforts to find these chemical intermediates proved unsuccessful. Among earlier hypotheses considering the nature of high-energy intermediates, was one based on the notion of specific protein conformation as the link between the energy-donating reaction of electron transport and the energy-accepting process of ATP formation [38].

After numerous futile attempts to identify $\tilde{X}$ with the special chemical intermediate, in 1961 Mitchell and Williams independently put forward the notion that ATP formation could be driven by protons but did not require the formation of any chemical intermediate [39–46]. This highly fruitful idea gave the strongest boost to the development of bioenergetics over the last 30 years. There was, however, an essential difference between Mitchell's and Williams' viewpoints on the role of a proton as the high-energy intermediate. This difference concerned not only the language used by both authors, but seems to be of principal importance in understanding the mechanism of membrane phosphorylation. Mitchell referred the protonmotive force to the transmembrane difference in electrochemical potentials of protons, $\Delta\bar{\mu}_{H^+}$, created between aqueous bulk phases separated by the coupling membrane [39, 40, 45, 46], The role of the bulk solvent necessarily implies the delocalization of energy released in the course of the energy-donating process (electron transport) over all surrounding spaces. Rejecting this point of view, Williams stated that "*protons in the membrane* rather than an *osmotic transmembrane gradient* of protons were required to drive the ATP synthesis" [43]. He referred to "the energies of *local* charges generated by the separation of protons from electrons by dislocated reactions".

As a matter of fact, the difference in understanding the role of protons in membrane phosphorylation reflects the existence of two different principal approaches to the problem of energy coupling in biomembranes that involve numerous concrete models. The first of them is most clearly expressed in the widely accepted Mitchell chemiosmotic concept [39, 40, 45, 46]. According to Mitchell's postulate, $\tilde{X}$ represents a transmembrane difference in the electrochemical potential of hydrogen ions, $\Delta\bar{\mu}_{H^+}$. This value can be presented in the form of the sum of two separately measured quantities: $\Delta\bar{\mu}_{H^+} = \Delta\varphi + (RT/F)\ln([H^+]_{in}/[H^+]_{out})$, where $\Delta\varphi = \varphi_{in} - \varphi_{out}$ is a transmembrane difference in electric potentials, and $(RT/F)\ln([H^+]_{in}/[H^+]_{out})$ is the difference in the chemical potentials of hydrogen ions on both sides of the coupling membrane that is determined by the difference in protons activities (concentrations). The expression for the protonmotive force $\Delta\bar{\mu}_{H^+}$ can be re-written as $\Delta\bar{\mu}_{H^+} = \Delta\varphi - 2.3(RT/F)\Delta pH$, where $\Delta pH = pH_{in} - pH_{out}$ is a transmembrane pH difference. The energy-donating electron transport processes eventually lead to the acidification and positive charging of chloroplasts intrathylakoid volume relative to the outer space (Figs. 5.1 and 5.2). In mitochondria the $\Delta\bar{\mu}_{H^+}$ has the opposite direction (the outer space becomes acidic and more positive than the mitochondrial matrix, Figs. 5.1 and 5.5).

Mitchell's original hemiosmotic hypothesis was founded on four main postulates [39, 40]. The author summarized them as follows (cited from [45]):

1. The ATPsynthase is a chemiosmotic membrane-located reversible protonmotive ATPase, having characteristic $\rightarrow H^+/P$ stoichiometry.
2. Respiratory and photoredox chains are chemiosmotic membrane-located protonmotive systems, having characteristic $\rightarrow H^+/2e^-$* stoichiometry, and having the same polarity of proton translocation across the membrane for normal forward redox activity as the ATPase has for ATP hydrolysis.
3. There are proton-linked (or hydroxyl ion-linked) solute porter systems for osmotic stabilization and metabolite transport.
4. Systems 1 and 3 are plugged through a topologically closed insulating membrane, called the coupling membrane that has a nonaqueous osmotic barrier phase of low permeability to solutes in general and to hydrogen ions and hydroxyl ions in particular.

These four fundamental postulates were the subject of experimental scrutiny during the last 30 years, and brought to light a great number of new facts and observations, and thus widely extended our knowledge in bioenergetics. The results of numerous experimental works carried out with mitochondria, bacteria, and chloroplasts were consistent with the main predictions of Mitchell's hemiosmotic hypothesis. It was not necessary to prove that the functioning of respiratory and photosynthetic redox chains could generate a

---

* In fact, this 1 : 2 stoichiometry was just a postulate which turned out to be correct (see below).

transmembrane proton gradient of electrochemical potentials. It should also be taken for granted that protonation/deprotonation reactions, associated with the electron transport processes, represent the obligatory intermediate steps in the chain of events that provide energy transformation in biomembranes. The functioning of the ETC complexes of energy-transducing organelles is associated with the vectorial transport of protons across their coupling membranes. There is no doubt that ATP formation in energy-transducing membranes is coupled with the protonation/deprotonation events in the H^+ATPases. Jagendorf's pioneering works on post-illumination phosphorylation in chloroplasts and ATP synthesis driven by an artificially imposed pH gradient (acid-base pH jump), were considered as among the first strongest experimental support for the chemiosmotic hypothesis. The acid-base transition in the Jagendorf–Uribe experiments artificially produces the Mitchell situation ($pH_{in} < pH_{out}$) without the normal electron transport.

Numerous foregoing investigations of the acid-base driven ATP synthesis in model-reconstructed systems were considered as paving the way to the final triumph of the Mitchell concept. It had originally been demonstrated in the classical works performed in Racker's and Kagawa's laboratories that the vesicles which contained only CF_0CF_1 complexes (without any electron transport components) were able to couple the formation of ATP from ADP and P_i with the flux of protons from vesicles artificially loaded with protons. Bacteriorhodopsin-driven proton pumps encrusted into the vesicles appeared to be competent in ATP synthesis by H^+ATPase complexes incorporated into the membranes of the same vesicles. External electric fields imposed on the coupling membranes of chloroplasts and mitochondria can lead to ATP formation from ADP and P_i. The reader can find a good deal of evidence concerning this subject in a number of original articles, reviews, and comprehensive monographs [13–15, 19–35, 45–47]. Here, we want to focus our attention on several biophysical aspects of hemiosmotic hypothesis that might be important from the viewpoint of the mechanism of coupling energy-donating and energy-accepting processes. For this reason, we reformulate the relevant points of chemiosmotic hypothesis as follows:

(i) The functioning of ETC generates the difference in the electrochemical potentials of protons across the coupling membrane forming topologically closed vesicles (mitochondria, thylakoids, and chromatophores).

(ii) Under phosphorylating conditions the value of the transmembrane difference of electrochemical potentials, $\Delta\bar{\mu}_{H^+}$, is high enough to ensure an energetically unprofitable reaction of ATP synthesis from ADP and P_i. This point, of course, implies that energetics of the ATP formation is considered from the position of conventional thermodynamics (see below).

(iii) The passage of protons down their electrochemical gradient through the ATPsynthase can lead to ATP formation irrespective of the manner of the $\Delta\bar{\mu}_{H^+}$ generation (ETC functioning, artificial creation of $\Delta\bar{\mu}_{H^+}$ by acid-base transition, or application of an electric field).

(iv) Both components of $\Delta\bar{\mu}_{H^+}$, $\Delta\varphi$ and ΔpH, are interchangeable and equally competent in driving ATP synthesis.

An orthodox version of the Mitchell hemiosmotic hypothesis assumes that both components, $\Delta\varphi$ and ΔpH, relate to the transmembrane differences between aqueous bulk phases separated by a coupling membrane. The decrease in proton free energy, associated with proton transfer through the H^+ATPsynthase along the electrochemical gradient, ensures the energy-accepting process of ATP formation. Thus, the hemiosmotic concept is founded on the traditional thermodynamic consideration of the balance between the changes in the Gibbs free energies of exoergic (discharge of $\Delta\bar{\mu}_{H^+}$ through the H^+ATPsynthase) and endoergic (ATP formation) processes. This means that a *"coupling device"* (H^+ATPsynthase) might at least be partly regarded as *"an entropic molecular machine"* which is capable of performing external work (or supporting the thermodynamically unfavorable process of ATP formation) *using the increase in entropy* in the course of a thermodynamically favorable process (the discharge of $\Delta\bar{\mu}_{H^+}$). As has been discussed in Chapters 2 and 3, the realization of endoergic chemical reactions occurs via accepting energy from the thermostat (macroscopic compartment) but not directly from the energy-donating processes.

Let us consider the total balance for an ATPsynthase reaction in biomembranes. According to the chemiosmotic concept, this reaction can be formulated as follows:

$$ADP + P_i + mH^+_{in} + nH^+ \rightleftarrows ATP + H_2O + mH^+_{out}. \qquad (5.2)$$

Here, n is the average number of protons bound to ATP due to the differences in the pK_a values of ATP, ADP, and P_i; in the presence of Mg^{2+}, $n \cong 1$ at pH 8, n decreases with lowering pH ($n \to 0$ at pH $\to$ 6). H^+_{in} and H^+_{out} symbolize "vector" protons transferred from the "positive" (P) to the "negative" (N) side of the coupling membrane, respectively. The coefficient m denotes the H^+/ATP stoichiometry for two coupled processes, the energy-accepting reaction of the ATP formation (5.3) and the energy-donating process (5.4) of proton translocation down their proton electrochemical potential. Let us divide (5.2) into two parts

$$ADP + P_i + nH^+ \rightleftarrows ATP + H_2O, \qquad (5.3)$$

$$mH^+_{in} \rightleftarrows mH^+_{out}. \qquad (5.4)$$

It follows from the Second Law of Thermodynamics that under steady state conditions the entropy production $\sigma = \sum J_j \mathscr{A}_j \geq 0$. In terms of the reaction (5.2) this means the fulfillment of the following condition $J_{ATP}\mathscr{A}_P + J_H\mathscr{A}_H \geq 0$, where J_{ATP} and J_H represent thermodynamic fluxes, i.e., the rates of ATP formation and coupled H^+ efflux, $\mathscr{A}_P$ and $\mathscr{A}_H$ are the corresponding thermodynamic forces. Here, $\mathscr{A}_P = \mathscr{A}_P^0 + RT \ln\{[([ADP][P_i])/[ATP]\}$ is the chemical affinity, or phosphate potential of the reaction (5.3), and $\mathscr{A}_H = m\Delta\bar{\mu}_{H^+}$, where $\Delta\bar{\mu}_{H^+} = \Delta\varphi - 2.3(RT/F)\Delta pH$. The coupling stoichiometry of the net process (5.2) should be determined by the condition $m = J_{H^+}/J_{ATP} \geq -\mathscr{A}_P/\Delta\bar{\mu}_{H^+}$. According to the majority of investigators who evaluated the stoichiometry of coupling by measuring the ratio $\mathscr{A}_P/\Delta\bar{\mu}_{H^+}$, $m = 3$ in mitochondria, submitochondrial particles, chloroplasts, chromatophores, and *E. coli* (for references see [13–14, 47–49]). Several authors, however, obtained controversial data indicating on the variable stoichiometry of coupling, $m \geq 3$–7 (for references, see [48–50]). It must be emphasized that during steady state phosphorylation, according to the laws of irreversible thermodynamics, a considerable part of the free energy transmitted from the energy-donating process to the energy-accepting process dissipates [50, 51].

If we consider the energy-donating process of a down-hill proton transfer $mH_{in}^+ \rightarrow mH_{out}^+$ as that occurring from one *delocalized* bulk phase to another, the energy released should be finally determined by the neutralization of certain basic groups, A_j^-, in the outer bulk phase (and/or spread over the outer surface of the coupling membrane):

$$mH_{out}^+ + \sum_{j=1}^{m} A_j^- \rightleftarrows \sum_{j=1}^{m} A_jH. \tag{5.5}$$

The coupling device using the drop in free energy that would occur as the result of a down-hill proton transfer, $mH_{in}^+ \rightarrow mH_{out}^+$, must operate as a typical *entropic machine*. It is an entropic term of free energy (concentration-dependent) that must determine its functioning. The energy released in the course of neutralizing these groups would quickly randomize over all degrees of freedom of a relatively large volume. Being delocalized in the bulk phase, the energy released cannot be efficiently transmitted directly to the active center of the coupling device.

This would not be the case, however, if it were possible to establish the mechanism of "local" or "direct" coupling. As we have mentioned above, according to Williams' viewpoint on the problem of energy coupling [41–44], the protons taking part in ATP formation are delivered directly to ATP-synthase without crossing the coupling membrane. In principle, this *local* mechanism does not require morphologically closed membrane vesicles. For this reason, the energy used for ATP synthesis was supposed to be kept in the form of "non-equilibrium localized charges" [43].

Another approach to the problem of energy transformation in biomembranes assumes that the coupling device operates as an *enthalpic molecular machine* which *directly* utilizes the energy released during the energy-donating process. For tightly coupled energy-donating and energy-accepting systems, the energy released has not enough time for dissipation over all degrees of freedom of the macroscopic compartment (thermostat), being at first localized predominantly on certain specific degrees of freedom. The relaxation concept of enzyme catalysis identifies such long-living states with the primary macroerg $\tilde{X}$ [52–54]. A fast local chemical change (the attachment of a ligand to the active center, ionization of an acid or basic group, etc.) is accompanied by the fast vibrational relaxation ($\tau \cong 10^{-10}$–10^{-12} s) of the active center and its immediate surroundings, while the structure of the whole macromolecular structure (e.g., the protein globule) remains practically unchanged. The subsequent changes are of a relaxation nature and might require correlated disruptions and the formation of a multitude of weak secondary bonds, i.e., the overcoming of a large entropy barrier. Conformational relaxation can last for microseconds, milliseconds, and even seconds. After fast chemical transformation of the enzyme active center, there appears a long-living out-of-equilibrium state: the active center is already changed, while the structure of the main volume of protein globule remains the same, i.e., initially the globule exists in an out-of-equilibrium state relative to the changed active center. There appears a structural strain between the relaxed and unchanged parts of the protein globule. This strain is slowly released in the course of protein conformational relaxation.

The basic ideas of this approach to the problem of energy coupling in biomembranes can be formulated as follows:

(i) Conformational relaxation of proteins and their complexes after fast local chemical changes should be regarded as motion along a specific mechanical degree of freedom.
(ii) This conformational relaxation is the working stroke of the corresponding molecular machine and is essentially an elementary act of the energy coupling of intracellular chemical reactions.

5.2. Transmembrane Electrochemical Proton Gradients in Chloroplasts

The main principles of membrane phosphorylation are the same in chloroplasts, mitochondria, and photosynthetic bacteria. In this section, in order to analyze the role of protonmotive force in the processes of energy transduction in biomembranes, we will focus our attention on the consideration of proton-transport processes in chloroplasts. In thylakoids the ΔpH is the main component of transmembrane difference in electrochemical potentials of hydrogen ions, $\Delta \bar{\mu}_{H^+} = \Delta \varphi - 2.3(RT/F)\Delta pH$. The conductivities of the thylakoid membrane for the majority of cations (Mg^{2+}, K^+, Na^+), existing in native chloroplasts or added to their suspension, are high as compared with that for hydrogen ions [12–14]. For this reason, the steady state transmembrane difference in electrical potentials is usually low, $\Delta \varphi \leq 20$ mV, and therefore the electrical component of $\Delta \bar{\mu}_{H^+}$ cannot be energetically competent for steady state ATP synthesis. Thus, the ΔpH component should be practically a sole contributor to $\Delta \bar{\mu}_{H^+}$ in chloroplasts under steady state conditions.

5.2.1. Brief Review of the Methods for the ΔpH Measurements with pH-Indicating Probes

There are two most frequently used methods for ΔpH determination in energy-transducing organelles with pH-indicating probes:

(i) the measurement of an energy-dependent spectral response of pH-sensitive indicators loaded into the vesicles [55–62]; and

(ii) the calculation of ΔpH values from the partition of permeable amines between the inner volume of the vesicles and the suspending medium [61–67].

Spectral methods for ΔpH measurements are based on the sensitivity of the absorption and fluorescence spectra, as well as on NMR [60] or EPR spectra, of the indicator molecules to the variations of the pH in the surrounding medium. The most frequently used pH-indicators are dye molecules whose optical and fluorescent spectra change with the protonation of indicatior molecules. The use of optical pH-indicators was fruitful in detecting protons evolving inside the thylakoids during the flash illumination of chloroplasts [55–59]. However, a serious problem could arise due to the interference with the nonspecific responses of dye redox reactions, and the superposition of optical changes from the electron transport components and electrochromic effects that take place during the illumination of chloroplasts with continuous actinic light [68, 69]. Many of the molecular indicators (e.g., 9-aminoacridine and neutral red [57]) bind to membranes, and this may be the reason why these indicators are not quite adequate probes for the determination of the transmembrane, i.e., the bulk phase-to-bulk phase, pH difference in energy-transducing organelles.

An amine distribution technique for the ΔpH measurements is based on the assumption that the partition of probing molecules between the vesicles' interior and the surrounding medium must be unambiguously determined by the ratio of hydrogen ion concentrations inside and outside the vesicles, $[A]_{in}/[A]_{out} = [H^+]_{in}/[H^+]_{out}$ (Fig. 5.6). Here, $[A]_{in}$ and $[A]_{out}$ are amine concentrations inside and outside the vesicles. The knowledge of inside and outside concentrations of the indicator enables us to calculate ΔpH =

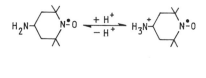

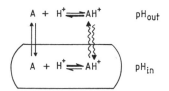

Fig. 5.6. Structural formulas of the neutral (A) and protonated (AH$^+$) forms of spin label TEMPOamine, and the scheme illustrating TEMPOamine partition between the external space and intravesicular volume.

$\log_{10}([A]_{in}/[A]_{out})$. With the acidification of the vesicle internal volume, the concentration of ΔpH indicating amines inside the vesicles would increase. Usually, an experimenter measures the total number, N_{in}, of indicator molecules absorbed by all vesicles in the suspension, but not the indicator concentration, $[A]_{in}$, inside the vesicles. In this case, the formula for the ΔpH calculation can be expressed in the following form:

$$\Delta pH = \log_{10}\left[(V/v) * \frac{Q}{1-Q}\right], \qquad (5.6)$$

where the values of v and V are the internal volume of the thylakoids and the total volume of the chloroplasts' suspension, and $Q = N_{in}/(N_{in} + N_{out})$ is the relative number of pH-indicating molecules inside the vesicles. This formula can be rewritten as

$$\Delta pH = \log_{10}([A]_{in}/[A]_{out}) - \log_{10}(1 - Q), \qquad (5.7)$$

where $[A]_{in} = N_{in}/v$ is the concentration of the indicator inside the vesicles, and $[A]_{out} = (N_{in} + N_{out})/V$ is the total concentration of the indicator added to the suspension.

To determine the ΔpH according to the commonly used formula (5.6), we need to know not only the number of indicator molecules inside the vesicles but also their internal volume, v. There is no great difficulty in finding the total internal volume of all kinds of vesicles in the suspension [63–65]. However, there could arise the insuperable problem of measuring the parameter v in the individual compartments in the *heterogeneous ensemble* of vesicles with different and/or variable internal volumes. For example, it is practically impossible to differentiate between the osmotic volumes of the grana- and stroma-exposed thylakoids in chloroplasts. To overcome this problem, a modified approach has been developed in our laboratory that allowed us to evaluate the ΔpH with a spin label TEMPOamine without knowledge of the internal volumes of the vesicles [70]. This method seems to be of special importance for measuring the ΔpH in certain compartments of the nonhomogeneous population of energy-transducing vesicles, such as, for example, the thylakoids of grana and stroma; it can also help to avoid the error in volume determination that might be caused by the energy-dependent osmotic effects. We will consider below the application of this method to the measurements of ΔpH in chloroplasts.

5.2.2. Measurements of ΔpH in the Thylakoids with the Kinetic Method

The original method for ΔpH evaluation in chloroplasts was proposed by Rumberg and Siggel in their pioneering work [71]. Having measured the dependence of the electron transport rate between two photosystems on the intrathylakoid pH_{in}, they were the first to determine the ΔpH value in spinach chloroplasts by measuring the reduction rate of the PS1 oxidized reaction center, P700$^+$. We will refer to this approach as the "kinetic" method.

The rate-limiting step in the chain of electron transport between two photosystems is the electron transfer from plastoquinol to a b_6/f complex. It has been demonstrated [72–75] that it is the plastoquinol oxidation step, but not its diffusion between spatially remote electron transfer complexes, that determines the rate of electron transfer between two photosystems. For a wide range of physiological temperatures and pH, the characteristic time of plastoquinol formation and its diffusion between PS2 and the cytochrome b_6/f complex is far less than the half-time of plastoquinol oxidation [74]. Plastoquinol oxidation is accompanied by releasing the protons into the thylakoid interior. This stage, therefore, should be controlled by the concentration of protons inside the thylakoids: the back pressure of hydrogen ions inside the lumen would hinder plastoquinol oxidation. The retardation of electron transport at the stage of plastoquinol oxidation can be easily detected as the slowing down of the rate of the reduction of the $P700^+$ oxidized reaction center of PS1. Thus, having the calibrating relationship between the rate of electron transport and the pH_{in} value in the lumen, we can determine the ΔpH from the kinetics data.

The kinetic method was used in our laboratory for measuring the transmembrane pH gradients in thylakoids under various metabolic states of chloroplasts [76, 77]. Let us consider these data in more detail. Figure 5.7 demonstrates the EPR spectra given by chloroplasts in the dark and under illumination by white light, and the kinetics of $P700^+$ reduction after a sudden switch-off of the light. Figure 5.8 shows the plots of the half-times, $\tau_{1/2}$, of the decay kinetics of $P700^+$ versus external pH values in uncoupled chloroplasts ($pH_{in} = pH_{out}$). This pH-dependence, used as the calibration curve for

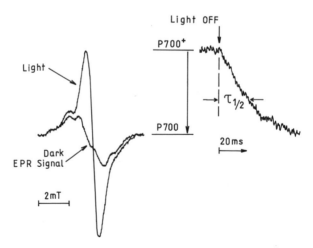

Fig. 5.7. EPR signals from isolated bean chloroplasts (left), and the kinetics of the light-induced EPR signal I decay after switching off the white light (right).

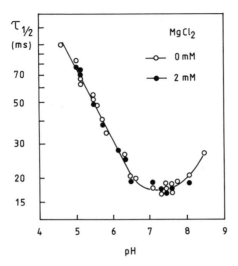

Fig. 5.8. The dependence of half-time, $\tau_{1/2}$, of P700$^+$ reduction (see Fig. 5.7) on the pH value of the chloroplast suspension, pH$_{out}$ (after [74]). Under experimental conditions used in [74] for measuring this dependence, the intrathylakoid pH$_{in} \cong$ pH$_{out}$.

the measurement of the internal pH$_{in}$, is one and the same in the presence of MgCl to the chloroplasts suspension as well as in Mg-depleted chloroplasts. According to [74], in the presence of Mg^{2+} ions the surface pH in the layer adjacent to the membrane, pH$_s$, should vary due to screening the membrane negative charges and the concomitant reduction in surface potential. The coincidence of both pH-dependencies can be regarded as evidence of the fact that it is a bulk phase pH$_{in}$ rather than a surface phase pH$_s$ that determines the rate of plastoquinol oxidation by the b_6/f complex. These data rule out the surface potential as the factor controlling electron transport between photosystems. Thus, the kinetic method can serve the role of the "pH-meter" sensitive to the pH value in the aqueous bulk phase of the thylakoid lumen.

Figure 5.9 shows plots of the half-time, $\tau_{1/2}$, of P700$^+$ reduction versus pH$_{out}$ for chloroplasts functioning under various metabolic states. Assuming that the rate of P700$^+$ reduction is controlled solely by the intrathylakoid pH$_{in}$, and using for the calibration curve the pH-dependence of uncoupled chloroplasts (+gramicidin, pH$_{in}$ = pH$_{out}$), we have determined the steady state ΔpH values established under phosphorylating conditions (state 3), or in the state of photosynthetic control (state 4). The data presented in Fig. 5.10 show that in state 4 (without added ADP), the ΔpH formed under steady state illumination increases from 1.6 to 3.2 with a rise in the external pH from 6 to 9.5. In state 3 (the excess of Mg^{2+}ADP in chloroplast suspension), the ΔpH value decreases reaching its minimal value ΔpH $\cong$ 0.5–1.0 at pH$_{out} \cong$ 7.5–8.0. Being associated with the efflux of protons from the thylakoids

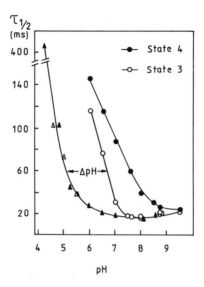

Fig. 5.9. The dependence of half-time, $\tau_{1/2}$, of P700$^+$ reduction after switching off the white light on the pH value of the chloroplast suspension. Light and dark symbols correspond to the chloroplast states 3 and 4, respectively, and triangles are uncoupled chloroplasts, $pH_{in} \cong pH_{out}$ (after [76]).

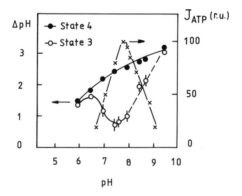

Fig. 5.10. Effects of the external pH on the ΔpH value determined by the kinetic method in states 3 and 4, and the rate of ATP formation ($- \times -$) (after [76, 77]).

through an actively functioning ATPsynthase complex, the minimal ΔpH value corresponds to maximal yields of ATP synthesis (Fig. 5.10).

5.2.3. Measurements of ΔpH in the Thylakoids with a Spin Labeling Technique

In order to compare, under similar experimental conditions, the results given by kinetic and conventional (use of ΔpH-indicating amines) methods, we turned to the EPR spectroscopy technique that enabled us to evaluate ΔpH by measuring the rate of electron transport between two photosystems [76, 77] or to calculate the ΔpH value from the partition of the spin label TEMPOamine [70, 78]. The results of our experiments were unexpected: both methods led to the same values of ΔpH under the state of photosynthetic control (state 4 according to [37], characterized by the lack of visible ATP synthesis), while there was a certain discrepancy in measuring ΔpH under phosphorylating conditions (an excess of added ADP). Analyzing these results [67, 77, 78], we concluded that the ambiguity in the ΔpH values measured by pH-indicating probes and kinetic methods could be explained by two reasons:

(i) the lateral heterogeneity of the chloroplasts' membranes; or
(ii) the existence of alternative pathways (transmembrane or intramembrane) for proton transport coupled with ATP synthesis.

Let us consider the results of our measurements of the ΔpH value in thylakoids with a spin label TEMPOamine [70, 78]. The EPR spectrum of TA in chloroplast suspension represents a triplet given by nitroxide radicals rapidly tumbling in polar surroundings (Fig. 5.11, spectrum a) with the rotational correlation time $\tau_c \cong 4 * 10^{-11}$ s. The EPR spectrum from TA molecules in the outer medium is broadened out by a charged paramagnetic compound (chromium oxalate), which is unable to penetrate inside the thylakoids, and thus cannot affect the EPR spectra from the minor portion of TA molecules located in the thylakoid lumen. The shape of the residual EPR signal from TA inside the thylakoids is slightly different from that of TA molecules in the external media (Fig. 5.11, spectrum b). A certain hindrance to TA motion inside the thylakoids ($\tau_c \cong 4.8 * 10^{-10}$ s) may be a consequence of the increased viscosity of structured water inside the thylakoids [78–80]. The value of the isotropic constant of hyperfine structure $a_{iso} = 0.167$ mT indicates that TA molecules inside the thylakoids are surrounded by polar media. There is also other independent evidence for TA localization in the bulk phase of the thylakoid lumen [66, 81]. In particular, it has been demonstrated that TA accumulates in the osmotic volume of the thylakoids. Therefore, we can conclude that TA is the probe for the measurement of a *transmembrane, i.e., the bulk phase-to-bulk phase, pH difference*. The ratio of the intensities of the residual signal, A_{in} (+ chromium oxalate), to the initial singal, A_0 (− chromium oxalate), is equal to the ratio of the internal, v, and

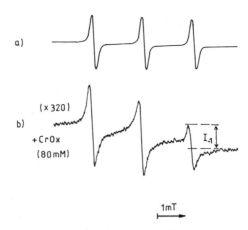

Fig. 5.11. The EPR spectra of spin label TEMPOamine in chloroplast suspension: (a) without broadening agents; and (b) in the presence of 80 mM chromium oxalate (after [78]).

the external, V, volumes, $v/V = A_{in}/A_0$. The ratio v/V obtained with the EPR method was in good agreement with similar estimations by other generally accepted methods [64, 65].

The uptake of TA in response to the light-induced acidification of the thylakoid lumen reveals itself as the increase in the EPR signal from TA molecules in the bulk phase of the intrathylakoid volume (Fig. 5.12, upper curve). This process is inhibited in the presence of the ionophore, i.e., gramicidin D (Fig. 5.12, lower curve).

If the light-induced uptake of TA leads to a high enough concentration of TA molecules inside the thylakoids, this would cause a concentrational broadening of the EPR spectrum. The calibration curve in Fig. 5.13 demonstrates that with increasing the TA concentration to 2.5 mM the amplitude of the EPR signal increases proportionally to the TA concentration, while it declines from the linear law as the TA concentration becomes higher than 2.5 mM. This deviation from the linear law, and a further drop in the signal amplitude, are caused by the effect of broadening the EPR spectrum due to more frequent collisions between the TA molecules at high concentrations. Thus, a high enough extent of the light-induced concentration of TA molecules inside the thylakoids could reveal itself by decreasing the height of the total EPR signal from TA in the chloroplast suspension. This effect makes it possible to detect the accumulation of TA molecules in the thylakoid lumen even in the absence of broadening compounds in an outer medium [78]. Figure 5.14 demonstrates the typical pattern of such a light-induced response to the EPR signal from TA in the chloroplast suspension. Concentrational broadening of the EPR spectra of TA molecules concentrating inside the thylakoids reveals itself as the reversible decrease in the EPR signal ampli-

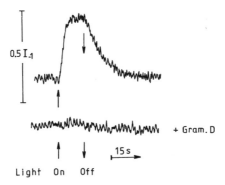

Fig. 5.12. The light-induced uptake of spin label TEMPOamine by the thylakoids revealed as the increase in the magnitude of the high field component of the EPR signal from TEMPOamine in a chloroplast suspension containing 16 mM chromium oxalate, the lower curve is in the presence of 20 μM gramicidin D (after [78]).

tude. Collapsing ΔpH after ceasing the illumination, the leak of TA molecules from the thylakoids is accompanied by narrowing the EPR line that is revealed in the rise of the EPR signal amplitude (parameter ΔA in Fig. 5.14).

The calibration curve (Fig. 5.13) shows that the broadening effect should manifest itself only if TA concentrations in the lumen, $[TA]_{in}$, becomes higher than the threshold level $[TA]_\theta = 2.5$ mM, i.e., if $[TA]_{in} \geq [TA]_\theta$. Thus, for every fixed value of ΔpH there should exist a critical concentration of TA in the suspension, $[TA]_0^*$, below which $[TA]_{in} \leq [TA]_\theta$, and thus the

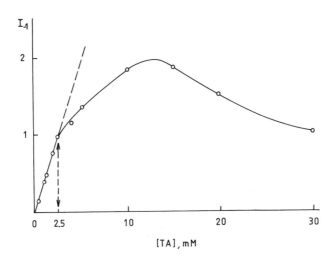

Fig. 5.13. The magnitude of the high field component of the EPR signal TEMPO-amine versus the TEMPOamine concentration (after [78]).

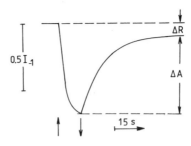

Fig. 5.14. The time-course of the light-induced changes in the magnitude of the high field component of the EPR signal from TEMPOamine in the suspension of bean chloroplasts (after [78]).

accumulation of TA molecules inside the vesicles would not cause the concentrational broadening of the EPR signal. This notion has been realized in [70] in order to determine ΔpH in grana- and stroma-exposed thylakoids functioning in various metabolic states.

Figure 5.15 demonstrates the dependencies of a TA reversible response (parameter ΔA) on TA concentration in a suspending medium, $[TA]_0$. We can now find the critical TA concentrations, $[TA]_0^*$, to be used for the calculation of ΔpH according to formula (5.7). If $Q \ll 1$ (this condition is fulfilled if the chloroplast concentration is not too high), then $\Delta pH \cong \log_{10}([TA]_\theta/[TA]_0^*)$. It is important to emphasize that this formula does not contain a value of the internal volume v. The values of $[TA]_\theta$ and $[TA]_0^*$ can be found from the calibration curve (Fig. 5.13) and from the concentrational dependence of the EPR spectra parameter ΔA (Fig. 5.15). The $[TA]_0^*$ value can be considered as that concentration of TA added to the chloroplast suspension which corresponds to the break point in the plot of the ΔA parameter versus $[TA]_0$. The broadening effect is absent at $[TA]_0 \le [TA]_0^*$ ($\Delta A \cong 0$), but we can see the rise in parameter ΔA with increasing TA concentration if $[TA] \ge [TA]_0^*$ (Fig. 5.15).

Using this method we have determined the pH transmembrane differences across the thylakoid membrane under various conditions of chloroplast functioning. In order to overcome the problem of structural and functional heterogeneity of thylakoid membranes, and thus to evaluate the ΔpH in grana- and stroma-exposed thylakoids in different metabolic states, we have investigated the light-induced uptake of protons and the paramagnetic pH-indicator (spin label TEMPOamine) by the thylakoids for two regimes of the chloroplasts' functioning:

(i) PS1-driven cyclic electron transport accompanied predominantly by the accumulation of protons inside stroma-exposed thylakoids; filling the proton pools of grana-exposed thylakoids, in this case, was minimized due to the inhibition of PS2-driven proton pumps.

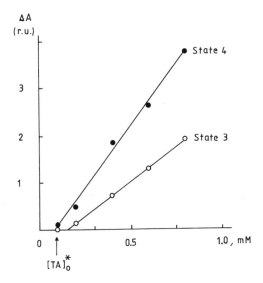

Fig. 5.15. The dependence of the magnitude on the reversible light-induced decrease of the high field component of the EPR signal from spin label on TEMPOamine concentration in the suspension of bean chloroplasts functioning under conditions of cyclic electron transport mediated by phenasine metosulphate, photosystem 2 was inhibited by diuron (after [70]).

(ii) Noncyclic electron flow driven by simultaneously collaborating PS1 and PS2; in this case, the proton pools of both kinds of thylakoids were effectively filled due to the active working of the proton pumps driven by the plastoquinol turnover shuttle and water-splitting complex.

The results of the measurements of the ΔpH (external pH 8.0) with the spin label TA under various conditions of chloroplast functioning are summarized in Table 5.1 [70, 82].

In the course of <u>noncylic electron transport</u> driven by simultaneously functioning PS1 and PS2, the values of ΔpH in the state of photosynthetic control (without added ADP, state 4) obtained with a spin label technique, $\Delta pH \cong 2.5$–2.9, are in a good agreement with that determined with other pH-indicating amines [63–67, 83–86]. Similar values of ΔpH were also ob-

Table 5.1.

Experimental conditions	Sorbitol	ΔpH (state 4)	ΔpH (state 3)
Noncyclic electron transport (PS1 + PS2),	10 mM	2.5–2.8	2.0–2.3
$H_2O \rightarrow$ methylviologen	800 mM	2.5–2.9	2.1–2.7
PMS-mediated cyclic electron transport (PS1)	10 mM	1.5–2.0	1.0–1.5
	800 mM	1.5–1.7	1.1–1.4

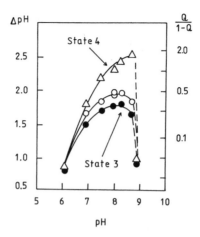

Fig. 5.16. The dependence of a steady state transmembrane pH difference, ΔpH, in bean chloroplasts on the pH value of chloroplast suspension (after [78]). ΔpH was determined from the partition of spin label TEMPOamine during noncyclic electron transport driven by photosystem 1 and photosystem 2 ($H_2O \rightarrow$ methylviologen). Triangles are without added ADP; light or dark circles are in the presence of 2 or 4 mM MgADP.

tained from studying the effects of various permeable buffers with different pK on the light-induced uptake of protons by the thylakoids [87]. There is also excellent agreement of these data with the ΔpH values in state 4 evaluated by the kinetic method (Fig. 5.16). Thus, all methods used for ΔpH measurements lead to the same results in the state of photosynthetic control.

This is not the general case, however, for chloroplast functioning upon phosphorylation conditions (an excess of added ADP, state 3). According to our determination of the ΔpH values with a spin label technique (Fig. 5.16, Table 5.1), in state 3, characterized by the intensive leak of protons via actively functioning ATPsynthase complexes, the steady state values of ΔpH were maintained at a high enough level, ΔpH $\cong$ 2.0–2.7 (depending on the osmomolarity of the chloroplast suspension). These values are lower, but only insignificantly, than that in state 4. Similar results were obtained with other ΔpH-indicating amines [83–86], as well as with the help of penetrating buffers [87]. On the other hand, the ΔpH measurement with the kinetic method [76, 77] indicates that the fast efflux of protons via functioning ATPsynthase precludes the generation of a significant pH gradient in state 3. According to kinetic measurements, ΔpH $\cong$ 0.5–1.5 (at external pH 7.5–8.0, corresponding to the highest rates of ATP synthesis, Fig. 5.10). Thus, ΔpH *under phosphorylating conditions appears to be dependent on the method used for the ΔpH measurement.*

Lower levels of the transmembrane pH gradients were registered with the spin label technique during PMS-mediated <u>cyclic electron transport</u> driven solely by PS1: $\Delta pH \cong 1.5$–2.0 in state 4 and $\Delta pH \cong 1.0$–1.5 in state 3 (at $pH_{out} \approx 8$). It is remarkable that such measured ΔpH values in state 3 correspond to those obtained from the kinetic data [76, 77].

According to [67, 77, 78, 87–88], the discrepancies in the ΔpH values may reflect the principal difference between two methods: kinetic measurements are sensitive to the pH values in certain locuses or *local* domains of the thylakoids, while the use of an amine distribution technique or pH-indicating dyes brings the information *averaged* over all compartments where pH-indicating molecules could be accumulated. There might be two possible reasons for the different ΔpH values given by the "local" and "averaging" methods:

(i) the nonuniform distribution of ΔpH along the laterally heterogeneous thylakoid membrane; and

(ii) the lack of equilibrium between the membrane-sequestered protonic pool and buffering groups in some regions of the intrathylakoid bulk phase.

If there is uniform distribution of the intravesicular pH over all compartments of the system, both methods would of course give the same results. This is probably the case for chloroplasts functioning in state 4. However, if there are compartments with different values of the internal pH (nonuniform distribution), the "local" and "averaging" methods could lead to quite different results (see, e.g., an example in Section 3.2.2).

Below we consider the evidence that the structural and functional heterogeneity of a chloroplast lamellar system may be the reason for different pH gradients across the membranes of grana- and stroma-exposed thylakoids.

5.2.4. Lateral Heterogeneity of ΔpH in Chloroplasts

The nonuniform distribution of electron transport and ATPsynthase complexes along the thylakoid membranes could lead to different pH gradients established across grana- and stroma-exposed thylakoid membranes. During noncyclic electron transport, driven by simultaneously collaborating PS2 and PS1, the proton pumps are effectively functioning in the thylakoids of both kinds. However, it is possible to imagine the situation with significant pH gradients in the thylakoids of grana- and relatively small ΔpH in stroma-exposed thylakoids [67, 78].

There are several lines of argument to support the notion of the nonuniform distribution of transmembrane pH gradients. One of them arises from the ΔpH measurements, while another one is from the comparison between the time-courses of proton accumulation inside the thylakoids during cyclic and noncyclic electron transport. The kinetic study of protons and TA uptake enables us to discriminate between the events associated with "loading" the protonic pools of grana- and stroma-exposed thylakoids. Comparing the uptake of protons and spin label TA by chloroplasts suspended in the media with different osmomolarity (and thus having different internal volumes), we

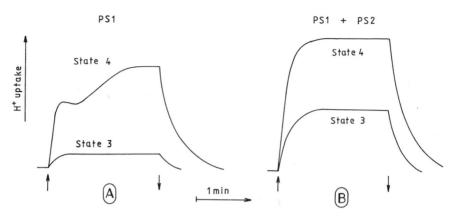

Fig. 5.17. The time-course of the light-induced pH changes in the suspension of bean chloroplasts functioning under cyclic (A) or noncyclic (B) electron transport conditions.

were able to calculate the number of protons binding to buffering groups located on (or buried inside) the membrane, as well as the number of protons bound to buffering molecules in the bulk phase of the lumen [87, 88].

During cyclic electron transport in state 4, chloroplasts suspended in the hypotonic medium reveal two-phase kinetics of the light-induced uptake of protons (as well as a pH-indicating spin label TA), while there are mono-phasic kinetics under phosphorylating conditions (Fig. 5.17(A)). A slow phase can be attributed to the additional accumulation of protons and TA inside the lumen of the swelling grana-exposed thylakoids. For chloroplasts functioning under the conditions of PMS-mediated cyclic electron transport the proton pumps of PS2 (most of them are located in the grana thylakoids) are not operating, and are thus inefficient in generating ΔpH in grana thylakoids. Only a small fraction of the PS1 and ATPsynthase complexes is located in the margins of the grana thylakoids. These PS1 complexes are relatively inefficient in generating the ΔpH in grana thylakoids in state 3. Meanwhile, in state 4, with the lack of an intensive efflux of protons via ATP-synthase, a gradual filling of the relatively large proton pools in the thylakoids of grana becomes possible. This reveals itself as the slow kinetic phase of the proton uptake (Fig. 5.17(A)). Such a slow phase is absent during the simultaneous functioning of both photosystems (PS1 + PS2), and the proton pools in the thylakoids of grana are filled faster (Fig. 5.17(B)) than in the case of cyclic electron transport (Fig. 5.17(A)).

The Model for the Partition of Protons inside Grana- and Stroma-Exposed Thylakoids. Figures 5.18 and 5.19 demonstrate sketches of the chloroplast cross section symbolizing filling up the proton pools for two kinds of compartments. We assume that these compartments are substantially different

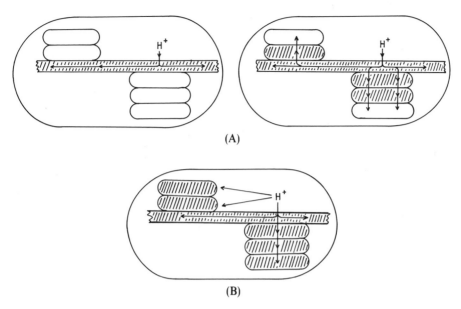

(A)

(B)

Fig. 5.18. Sketches illustrating the light-induced filling of the proton pools of grana- and stroma-exposed thylakoids in the state of photosynthetic control: (A) PS1-catalyzed cyclic electron transport; and (B) noncyclic (PS1 + PS2) electron transport.

with respect to the content of the electron transport and ATPsynthase complexes: compartments of the first kind contain predominantly PS1 and ATP-synthase complexes, while the other one contains only PS2 complexes. According to [89–98], stroma-exposed membranes contain about 70–90% of $CF_1 CF_0$ and 75% of PS1 complexes, while about 90% of PS2 complexes are located in the region of pressed grana membranes.

Figure 5.18 presents the sketches for the events that take place in state 4.

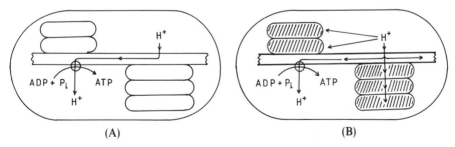

(A)   (B)

Fig. 5.19. Sketches illustrating the light-induced filling the proton pools of grana- and stroma-exposed thylakoids under photophosphorylation conditions. (A) PS1-catalyzed cyclic electron transport; and (B) noncyclic (PS1 + PS2) electron transport.

In the course of PS1-driven cyclic electron transport (Fig. 5.18(A)), initially there occurs a predominant filling of the proton pools in stroma thylakoids. This stage corresponds to the first phase in the kinetics of proton uptake. Being remote from the thylakoids of grana, the major part of the PS1-driven proton pumps are inefficient in filling the proton pools of grana-exposed thylakoids. The attainment of proton equilibrium between both types of thylakoids proceeds only slowly. This equilibrium could be finally achieved due to the functioning of a minor (marginal) part of PS1 complexes located in grana, as well as to the relatively slow diffusion of protons from the stroma to grana intrathylakoid volume. Just these processes are probably responsible for the occurrence of the second (slow) phase in the kinetics of proton uptake (Fig. 5.17(A)). The extent of this phase depends on the internal buffering capacity of the thylakoids, and can be varied by incubating the thylakoids in the media with different osmomolarity and concentrations of the permeable buffers. On the other hand, during noncyclic electron transport (Fig. 5.18(B)), the proton pools of grana thylakoids are filled immediately due to the functioning PS2 complexes; this provides the faster attainment of protonic equilibrium than in the case of the PS1-driven cyclic electron transport.

Figure 5.19 symbolizes the protonic events inside the thylakoids that occur under phosphorylating conditions. During PS1-driven cyclic electron transport (Fig. 5.19(A)), the greater part of the protons pumped into the lumen of stroma-exposed thylakoids would not have enough time to reach the lumen of grana-exposed thylakoids. This would be the case if the turnover time of ATPsynthase ($\tau_{1/2} \cong 2.5$ ms) is less than the characteristic time of proton diffusion. Thus, the competition between the two processes, a fast coupled efflux of protons via ATPsynthase located mainly in stroma-exposed thylakoids and a relatively slow deposition of protons from stroma- to grana-exposed thylakoids, would preclude the filling of grana proton pools by PS1-driven proton pumps (Fig. 5.19(A)). On the other hand, during noncyclic electron transport functioning, PS2 complexes could support the relatively high level of ΔpH across the membranes of grana-exposed thylakoids, thus masking the low level of the ΔpH in stroma-exposed thylakoids (Fig. 5.19(B)). Meanwhile, a hindrance to the intrathylakoid diffusion of protons from the lumen of grana to the lumen of stroma would not compensate for the fast efflux of protons from the actively phosphorylating thylakoids of stroma. Therefore, in state 3 the ΔpH in stroma thylakoids could be less than that in grana thylakaids (Fig. 5.19(B)).

Thus, there may be two possible reasons for the different steady state levels of ΔpH in these compartments:

(i) a more intensive efflux of protons via functioning ATPsynthase from stroma-exposed than from grana-exposed thylakoids; and
(ii) the hindrance to the long-range diffusion of protons from the lumen of grana-exposed to the lumen of stroma-exposed thylakoids that would preclude the randomization of proton gradients over the lumens of both kinds of thylakoids.

The first reason, as we discussed above, is a direct consequence of the thylakoid membrane lateral heterogeneity, characterized by dominating the ATPsynthases and PS1 complexes in stroma-exposed thylakoids. It is probably the heterogeneity of thylakoids that determines the different effectiveness of PS1 and PS2 for ATP synthesis that has been demonstrated in de Kouchkovsky's and Ort's laboratories [99–101]. A comparative study of the coupling efficiency during the two modes of chloroplast functioning revealed that protons pumped by PS2 into the lumen of grana-exposed thylakoids were less efficient in photophosphorylation than protons pumped by PS1. According to [101], the flash-induced ATP formation driven by PS2 operating alone required nearly twice as many flashes as were necessary when PS1 and b_6/f complexes operated together.

The second reason for the nonuniform distribution of ΔpH is that of the diffusion barriers resisting lateral proton transfer, preventing proton randomization inside the lumens of grana- and stroma-exposed thylakoids. This barrier should hinder the establishment of equilibrium between proton pools of these thylakoids. This situation could be realized if the characteristic time for the long-range diffusion of hydrogen ions from actively functioning PS2 complexes (located in grana-exposed thylakoids) to remote CF_1CF_0 complexes (situated in stroma-exposed thylakoids) was much greater than the characteristic time of the proton efflux from the lumen of stroma via actuating ATPsynthase. According to the measurements of Graber and his colleagues [32–35], the turnover time of the chloroplast ATPsynthase is about 2.5 ms. This time is less than the characteristic time of the rate-limiting step of the noncyclic electron transport, i.e., the protolytic reaction of plastoquinol oxidation by a b_6/f complex ($\tau_{1/2} \cong 20$ ms [102–104]). This means that the efflux of protons through the functioning ATPsynthase could occur before the next proton from plastoquinol would be injected into the lumen of stroma-exposed thylakoids. Thus, a relatively fast efflux of protons would hinder the building of a proton gradient across stroma-exposed thylakoid membranes.

Under certain conditions, the coupled efflux of protons will not be fully compensated for by the fast turnover of remote PS2-driven proton pumps. This would be the case when ATPsynthase turnover occurs faster than the long-distance diffusion of protons from the lumen of grana-exposed thylakoids to ATPsynthase complexes located in stroma-exposed thylakoids. It is well known that the effective coefficient for proton diffusion in water solutions is reduced by the presence of membrane-bound buffering groups [105–107]. Thus, the rate of lateral proton diffusion in the lumen may be rather slow. If this is true, protons intensively pumped by PS2 inside grana thylakoids would not have enough time to be readily equilibrated with the proton pools in stroma-exposed thylakoids. Thus, lateral heterogeneity and diffusion barriers could preclude randomizing the concentrations of protons inside grana- and stroma-exposed thylakoids. The situation is similar to that described by Ort [108]: "Hydrogen ions entering the inner aqueous phase no longer belong kinetically to the pool of protons participating in ATP formation." The conclusion regarding the existence of the lateral gradient of ΔpH, has been reached by de Kouchkovsky and his colleagues who investigated the isotopic effects in chloroplasts [109].

The coexistence of the thylakoids with different ΔpH can manifest itself as the discrepancy between the ΔpH values measured by two different methods, i.e., kinetic and pH-probing amines. *The kinetic method is equivalent to the use of a "local pH-meter,"* which gives information about ΔpH in certain local compartments of chloroplasts, while ΔpH-*indicating amines give information averaged out at all kinds of chloroplast compartments.* The latter method, being a "nonlocal (or averaging) pH-meter," cannot discriminate the ΔpH in compartments with different internal pH_{in}. If the essential part of pH-sensitive electron transport complexes is localized in stroma-exposed thylakoids with relatively small pH gradients, then just these thylakoids would determine the value of ΔpH registered by the kinetic method. Meanwhile, pH-indicating amines, accumulating in grana-exposed thylakoids, would still detect rather high values of ΔpH created in grana by the effective functioning of PS2 proton pumps, masking the low ΔpH in other compartments. As we stressed in Chapter 3, the "*average*" ΔpH may be significantly different from the real ΔpH in certain functionally important compartments or in the vicinity of ATPsynthase. Thus, the use of averaged ΔpH has no more value than making a diagnosis of a certain person using "average temperature" or "average blood pressure" formally calculated for all patients in a hospital.

The comparison of ΔpH values, generated during cyclic and noncyclic electron transport (Table 5.1), supports the idea of a nonuniform ΔpH distribution along the chloroplast lamellas. We were able to evaluate the ΔpH in stroma-exposed thylakoids with a TA spin label, avoiding significant overestimation of the ΔpH value that might be due to the interference with the high ΔpH value in grana-exposed thylakoids. For PMS-mediated cyclic electron transport driven by PS1 (with DCMU-inhibited PS2), there is no substantial filling of the proton pools of the grana-exposed thylakoids, and therefore the thylakoids of grana do not accumulate significant "masking" amounts of TA molecules. In this case the discrepancy disappears between the results of ΔpH measurements by the kinetic and spin labeling methods. The use of TA demonstrates that during cyclic electron transport in state 3 chloroplasts generate a relatively small transmembrane pH gradient (Table 5.1); the values ΔpH $\leq$ 1.0–1.5 at $pH_{out} \approx 8$ are in a good agreement with ΔpH values measured with the kinetic method under phosphorylating conditions (Fig. 5.10). These results imply that *the kinetic method is equivalent to the use of the local pH-meter,* being sensitive mainly to the internal pH in stroma-exposed thylakoids enriched with PS1 complexes.

Both methods, the kinetic and spin labeling techniques (the latter corresponds to the use of the conventional amines distribution method), demonstrate that steady state phosphorylation in chloroplasts can occur at relatively low transmembrane pH deference, ΔpH $\cong$ 1.0–1.5. Recently we were glad to find independent support: the same values of the ΔpH have been pub-

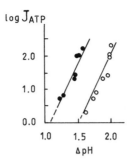

Fig. 5.20. The rate of ATP formation versus the ΔpH value in the thylakoids (redrawn from [110] with permission).

lished by Dilley's group [110]. These authors used the $[^{14}C]$methylamine distribution technique for measuring the steady state ΔpH values in pea chloroplasts. Figure 5.20 reproduces the results of their experiments, demonstrating the flux–force relationships (the rate of ATP synthesis versus the ΔpH value) that were obtained by varying the light intensity. These data clearly demonstrate that steady state ATP formation can proceed at a high rate under rather low pH gradients, characterized by the threshold levels: ΔpH$_{th}$ = 1.1 and ΔpH$_{th}$ = 1.5, for high- and low-salt stored chloroplasts, respectively. Similar results were obtained in Dilley's laboratory with the use of a pH-sensitive fluorescent dye (R.A. Dilley, personal communication). These results are in good accord with our results of ΔpH measurements for chloroplasts suspended in low- or high-osmotic media (Table 5.1). All the data presented above demonstrate that *effective photophosphorylation in chloroplasts can proceed under a relatively small transmembrane pH difference across the thylakoid membranes*, ΔpH $\leq$ 1.0–1.5.

In thylakoids the ΔpH value is the main term of transmembrane difference in the electrochemical potentials of hydrogen ions, $\Delta\bar{\mu}_{H^+} = \Delta\varphi + 2.3(RT/F)|\Delta pH|$. The conductivities of the thylakoid membrane for the majority of cations (Mg^{2+}, K^+, Na^+), existing in native chloroplasts or added to their suspension, are high as compared with that for hydrogen ions (see, for references, [111]). For this reason, the steady state transmembrane difference in the electrical potentials is usually low, $\Delta\varphi \leq$ 20 mV. According to [112], under continuous saturating illumination the transmembrane potential difference across the thylakoid membrane decreases from 80 mM to 0 mM by increasing the KCl concentration in the suspending medium from 1 mM to 100 mM. Therefore, the electrical component of $\Delta\bar{\mu}_{H^+}$ cannot be energetically competent for steady state ATP synthesis under normal conditions. Thus, $\Delta\varphi$ cannot compensate relatively low ΔpH values in steady state 3 which should be practically the sole contributor to $\Delta\bar{\mu}_{H^+}$. High rates of steady state ATP synthesis at relatively low ΔpH values might indicate the existence of a local mechanism of energy coupling in chloroplasts. Meanwhile, the transmembrane (bulk phase-to-bulk phase) pH difference can play a crucial role, thus providing conditions that are obligatory to ensure the reiterative mode of ATPsynthase functioning under steady state or quasi-stationary conditions (for more detail, see Section 5.3.2).

5.2.5. Membrane-Sequestered Proton Pools and Alternative Pathways of Proton Transport Coupled with ATP Synthesis

The orthodox version of the Mitchell chemiosmotic hypothesis assumes that the transmembrane difference in proton electrochemical potentials, $\Delta\bar{\mu}_{H^+}$, is the direct driving force ensuring the energy-accepting process of ATP formation. This point of view implies that protons injected or translocated into the thylakoids are readily randomized over the whole lumen's volume. The internal bulk phase represents the common reservoir for protons pumped inside due to energy-donating reactions and protons effluxed through H^+ATPsynthase complexes. This teaching was questioned by several investigators who suggested the so-called *"local"* mechanisms of energy coupling. According to Williams' idea [41–44], the conditions for ATP formation are provided due to increasing the activity of protons in a certain locus within an energy-transducing membrane. This and similar foregoing approaches to the problem of energy coupling in biomembranes suggest that the proton pumps inject the protons into membrane domains and deliver them immediately to ATPsynthase via a lateral intramembrane pathway (see, for references, [113]), or along the membrane interface [114], thus bypassing the bulk phase. Such an approach does not imply proton equilibrium between the proton-accepting groups that belong to the membrane proton-conducting pathway and proton-buffering groups located in the aqueous bulk phase of the thylakoid lumen. If this were true, the transmembrane difference in proton electrochemical potentials, $\Delta\bar{\mu}_{H^+}$, cannot be considered as the immediate source of energy for ATP formation.

The main line of argument in support of the "local" coupling mechanism arises from the measurements of the force–flux relationships in energy-transducing organelles. An obvious deficit in the $\Delta\bar{\mu}_{H^+}$ value related to the bulk phases is usually considered as evidence of the local mechanism. One example of such kinds of an apparent energy deficit in mitochondria is presented in the results obtained by Wilson and Forman [115]. Their measurements of the $\Delta\varphi$ and ΔpH values were not consistent with the transmembrane pH gradient across the mitochondrial membrane playing the role of an energy source for ATP synthesis. The ΔpH did not appear to be energetically coupled to the electrical potential difference, $\Delta\varphi$. At a constant rate of electron transport and phosphate potential, the decreasing ΔpH value was not accompanied by an energetically equivalent rise in the $\Delta\varphi$ value, i.e., the ΔpH and $\Delta\varphi$ terms did not reveal the interchangeability that had been postulated by hemiosmotic hypothesis. The stoichiometric ratio, $\Delta G_{ATP}/\Delta G_{H^+}$, ranged from 2.8 to 7.0 depending upon experimental conditions. The authors have concluded it was unlikely that proton transport is equilibrated with the high-energy intermediates of oxidative phosphorylation. Other examples of variable stoichiometry, as well as the numerous arguments *pro* and *contra* the localized or delocalized mechanism of energy coupling, were scrutinized at length in several original and reviewing articles (for references see [47–50, 113, 114, 116–117]). Detailed analysis of this problem is beyond the scope of this book; below we will focus our attention only on several works concerning proton transfer in chloroplasts.

There were several experimental indications of existing alternative ways for the proton transfer to an active center of the chloroplast's ATPsynthase: transmembrane (from the *delocalized* internal bulk phase of the lumen to the external bulk phase) and lateral (*localized* in membrane domains and including the proton translocation along the membrane proton-conducting path, or its diffusion along the membrane interface, Fig. 5.21). Experiments with permeating buffers have demonstrated that the pathway of protons from the electron transport complex to ATPsynthase does not involve the whole aqueous bulk phase of the intrathylakoid volume [87, 88, 110, 113, 118–126]. There are indications that both methods of proton transfer, transmembrane (hemiosmotic) and lateral (nonhemiosmotic), may be involved in the processes of ATP synthesis: under certain conditions one protonic pathway is converted to another one. In particular, the data obtained by Dilley's group [121–122] led to the conclusion that in chloroplasts the switching from one regime to another is salt-controlled: the lumen bulk phase "does not participate as part of the proton diffusion pathway" in KCl-washed chloroplasts (nonhemiosmotic mode of chloroplast functioning), while the KCl-treatment reverts the chloroplasts to the chemiosmotic mode of proton transfer. This conclusion has been drawn from the study of the influence of permeable buffers that can increase the buffering capacity of the lumen's bulk phase on the lag phase in the kinetics of ATP formation.

A similar conclusion follows from the work of de Kouchkovsky's group who measured the rate of ATP formation driven by a cyclic (PS1) or non-cyclic (PS1 + PS2) electron transport chain as a function of the apparent ΔpH value [99]. The hypothesis of the direct interaction of PS2 protonic domains with ATPsynthase was also considered by Yaguzhinsky's group [124] who studied the effects of the uncoupler gramicidin D on the photophosphorylation rate in chloroplasts during cyclic or noncyclic electron transport.

A notion that the proton-buffering domains sequestered in thylakoid membranes may be involved in proton diffusion has been elaborated on in several laboratories [59, 113, 125–130]. There is much data indicating the existence of membrane-bound proton pools which are not in rapidly established equilibrium with protons in the bulk phases (see, for references, [113]). These membrane-buried slow exchanging proton-accepting groups belong to the thylakoid membrane proteins. Under normal conditions, the intramem-

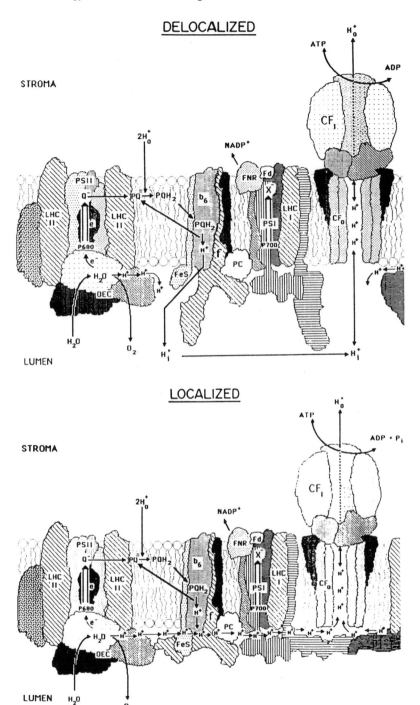

Fig. 5.21. "Delocalized" and "localized" models of proton transfer coupled with ATP synthesis in chloroplasts. (Reproduced, with permission, from the *Annual Review of Plant Physiology*, Volume 38, © 1987, by Annual Reviews Inc.)

brane proton domains are metastable. Such domains are separated from the aqueous bulk phases by low-permeable barriers. The addition of protonophores or a certain treatment of chloroplasts (mild heating, removal of the CF_1 factor, etc.) lower the barriers, and thus hasten the loss of protons from the sequestered buffering groups to the bulk phases.

Using acetic anhydride as a probe for the sequestered proton-buffering domains, Dilley and his collaborators were able to suggest that these domains involve buffering groups with pK $\cong$ 7.8 that originate from certain lysin residues in a positively charged environment [125]. Such buried lysin groups were attributed to three proteins of the water-splitting complex in PS2, Cytochrome f, light-harvesting pigment proteins [113]. The most interesting fact is the identification of one such lysin residue as Lys 48 of the 8-kD subunit of the membrane portion of ATPsynthase. It might be reasonable to assume that it is this group that is directly involved in the membrane-sequestered proton-conducting path and participates in the protonation/deprotonation events in the CF_0–CF_1 complex.

Filling sequestered protonic pools occurs in the course of chloroplast electron transport functioning. Using a lipid soluble pH-indicating dye, neutral red, as a sensor for protons, Junge and collaborators investigated the deposition of protons into the thylakoid lumen [59, 129]. They determined that under certain circumstances the protons evolved by PS2 are trapped by membrane-sequestered proton-accepting groups, thus escaping detection by neutral red. "Missed" protons were not detected, either in the lumen or in the outer bulk phase. Only after the saturation of sequestered protonic domains were the protons released by PS2 detected in the thylakoid's lumen. Meanwhile, the slow component in the kinetics of the neutral red response ($\tau_{1/2}$ = 20 ms), related to PS1-driven electron transport, was not dependent on the protonation/deprotonation of membrane domains.

As we discussed above, high rates of steady state ATP synthesis at relatively low transmembrane $\Delta\bar{\mu}_{H^+}$ values should indicate the existence of a local mechanism of energy coupling in chloroplasts. This mechanism implies, in particular, that the proton transfer to the CF_1 active center is by some way other than the transmembrane one. The existence of an intramembrane (or interface) protonic pathway suggests an alternative explanation of our results concerning the ΔpH measurements from the kinetic data. If the protons evolved in the course of plastoquinol oxidation do not readily equilibrate with the buffering groups in the internal bulk phase, the kinetic method could ignore the rather high transmembrane pH difference detected by the pH-indicating amines accumulating in the lumen. This situation might be realized if the protons released are quickly delivered to ATPsynthase along the membrane interface or through the membrane-sequestered pathway. If this is the case, the *transmembrane* pH *difference cannot be considered as the* *immediate* *source of energy driving the elementary act of* ATP *synthesis.* Meanwhile, the transmembrane (bulk phase-to-bulk phase) pH difference can play a crucial role, thus providing conditions that are obligatory for ensuring the reiterative mode of ATPsynthase functioning under steady state or quasi-stationary conditions [82, 190].

5.3. Mechanism of ATP Formation Catalyzed by H^+ ATPsynthases

5.3.1. An Elementary Act of ATP Synthesis

5.3.1.1. Initial Events of ATP Formation

The soluble portion of ATPsynthase (coupling factor F_1) of most kinds of energy-transducing membranes is composed of six large subunits (3α and 3β with molecular masses of about 59 kDa and 56 kDa, respectively), and three smaller subunits, designated as γ, δ, and ε (with molecular masses of 37 kDa, 17.5 kDa, and 13 kDa, respectively). Figure 5.22 depicts a tentative scheme of the ATPsynthase arrangement in the membrane (for references concerning the structural and functional properties of F_0F_1 complexes, see the recent reviews [131–146]). All α and β subunits contain nucleotide binding sites. The total number of nucleotide binding sites on F_1 (for ADP or ATP) is six. Three of them are located on the noncatalytic centers of α subunits. Coupling factors isolated from various sources usually contain one to four tightly bound adenine nucleotides.

There are two types of ATP bound to F_1: tightly bound but exchangeable ATP molecules that represent catalytically active intermediates (or "transitory" bound ATP), and practically nonexchangeable tightly bound ATP. In the course of enzyme turnover, nucleotides bound to noncatalytic centers of α subunits do not escape from the enzyme molecule, although ATPsynthase

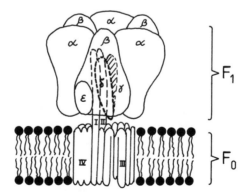

Fig. 5.22. Schematic representation of the F_0F_1 complex in a coupling membrane (drawn on the basis of work from [208]).

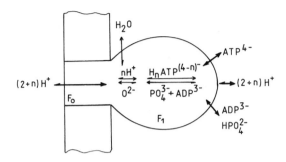

$$2H^+ + \ ^-O-P\overset{O^-}{\underset{O}{\lessgtr}}O^- + \ ^-OADP \longleftrightarrow H^+ + HO\overset{^-O}{\underset{O}{\lessgtr}}P\overset{O^-}{\lessgtr}OADP \longleftrightarrow H_2O + \overset{^-O}{\underset{O}{\lessgtr}}P-OADP$$

Fig. 5.23. Diagram illustrating the Mitchell mechanism of direct chemiosmotic coupling (after [46, 153]).

functioning promotes a very slow exchange of nucleotides bound to the noncatalytic sites. Three exchangeable sites of nucleotide binding are located at the catalytic centers of β subunits. Most of the tightly bound ADP molecules attached to β subunits are involved in ATP formation.

All the data available point to ADP as the immediate acceptor of P_i. There are no other phosphorylated precursors other than ADP, thus, ATP is the first compound formed from P_i in the course of photophosphorylation in chloroplasts and chromatophores, and oxidative phosphorylation in mitochondria. The data obtained in several laboratories demonstrated that adenine nucleotides bound to F_1 can incorporate $^{32}P_i$ from the medium, forming $[^{32}P_i]$ATP labeled immediately in the γ position [147–151].

The chemical aspects of ATP formation and hydrolysis catalyzed by H$^+$ATPases have been investigated by several authors [152–159]. Mitchell suggested that ADP phosphorylation proceeds as the result of an "in-line" nucleophylic displacement mechanism, as shown in Fig. 5.23. According to this mechanism, protons are directly involved in ATP synthesis, attacking the O$^-$ group of inorganic phosphate. This process is accompanied by a concerted attack on P_i by the appropriate O$^-$ group of the ADP molecule. The negative charges on the interacting oxygen atoms were believed to be neutralized by the appropriate surrounding in the catalytic domain of F_1. Bearing in mind the balance of the electric charges of the reagents depicted in Fig. 5.23, Mitchell concluded that the vectorial disposition of the oxide ion (O^{2-}, in Mitchell's terminology) from P_i is equivalent to the translocation of two protons in the opposite direction. According to this mechanism, the withdrawal of O^{2-} from P_i, associated with the reversible ATP formation, should be facilitated by the translocation of two protons to the F_1 active center. As

Mitchell proposed, the osmotic protonmotive force may be applied directly to the hydroxide or "oxide ion," pulling it into the catalytic center of F_1.

Important features of the initial events of ATP synthesis were clarified following the measurement of the time-course of ATP formation in chloroplasts with the use of rapid mixing and quenching techniques. Smith and Boyer [22] had demonstrated that energization of the thylakoid membranes, as a result of the acid-base transition, led to the formation of tightly bound ATP from ADP and P_i. The incorporation of $^{32}P_i$ in ATP initiated by the pH jump reveals a lag-phase of only about 3–7 ms (at 25 °C) [23]. This newly formed tightly bound $[^{32}P]$ATP molecule represents a transient intermediate in the catalytic sequence of ATP synthesis. Most of the ADP and P_i bound at a catalytic site of CF_1 is committed to ATP synthesis without interchanging with ADP and P_i in the medium. However, a very important observation was made in Boyer's laboratory [159]. The authors had demonstrated that ADP molecules from the medium can be involved in the initial steps of acid-base induced ATP formation in chloroplasts. Subsequent release of tightly bound nucleotides occurs within the turnover time of the enzyme, $\tau \leq 10$ ms. The authors proposed that an energy-requiring conformational transition in CF_1 "loosens ATP binding at one site and promotes P_i and ADP binding at the other site." Rosen et al. [160] demonstrated that such an "intermediate" form of tightly bound ATP was also formed under normal conditions, i.e., during steady state ATP synthesis in chloroplasts. These "intermediate" ATP molecules may remain bound at the catalytic center of CF_1 after ceasing energization of the thylakoid membranes. During steady state phosphorylation, at a given moment of time, most or all phosphorylation complexes are functional, i.e., more than one catalytic center per F_1 molecule participates in ATP synthesis.

5.3.1.2. Energy-Requiring Step of ATP Formation

In 1973–1974 two independent groups of scientists, from the laboratories of Boyer [161, 162] and Slater [150, 151, 163], originally suggested that energy inputted to ATPsynthase must be used mainly to promote the liberation of a tightly bound ATP from the F_1 catalytic center. Boyer's group reached this conclusion from the results of experiments on ATP–H^{18}OH and P_i–H^{18}OH isotopic exchange, while the arguments of Slater's group were based on studying the energy-dependent changes in nucleotide binding. This idea became a prominent milestone on the way to understanding the intimate mechanism of ATP synthesis.

In aqueous solution the reaction of ATP formation from ADP and P_i is unprofitable for any reasonable concentration of ATP, ADP, and P_i in a cell. Within the pH interval from 6 to 9, and Mg^{2+} concentrations up to 50 mM, the equilibrium constant, K_{eq}, of the reaction ADPH^{2-} + H$_2$PO$_4^-$ $\rightleftarrows$ ATPH^{3-} + H$_2$O (this equation is written, for certainty, just for one state of the reagent's ionization, while there are other possible states depending on the pH and salt concentrations) was determined to be confined within the limits $K_{eq} \cong (0.02\text{--}25) * 10^{-6}$ [164]. These values of the equilibrium constant correspond to the standard affinity $\mathscr{A}^0 \cong -30$ to -45 kJ/mole. This means that for any reasonable ranges of ATP, ADP, and P_i concentrations the equilibrium of the reaction $ADPH^{2-} + H_2PO_4^- \rightleftarrows ATPH^{3-} + H_2O$ should be shifted to the left side, and, therefore, the equilibrium ratio of ATP and ADP concentrations in the solution must obey the condition $[ATP]_s/[ADP]_s \ll 1$. However, the ratio of ATP and ADP bound to F_1 can significantly differ from the ratio of their equilibrium concentrations in aqueous solution.

Let us consider the simplest formal scheme of the ATPsynthase reaction

$$E + ADP + P_i \underset{k_{-1}}{\overset{k_1}{\rightleftarrows}} \underset{(1)}{E \cdot ADP \cdot P_i} \underset{k_{-2}}{\overset{k_2}{\rightleftarrows}} \underset{(2)}{E \cdot ATP} \underset{k_{-3}}{\overset{k_3}{\rightleftarrows}} \underset{(3)}{E + ATP.} \quad (5.8)$$

with H_2O entering at step k_2 and H_2O at step k_{-2}.

For the sequence of processes (5.8) the equilibrium constant, K_{eq}, related to the reagents in aqueous solution, can be expressed using the Haldane relationship

$$K_{eq} = (k_1 k_2 k_3)/(k_{-1} k_{-2} k_{-3}) = K_{eq}^*(K_{ATP}/K_{ADP,P_i}). \quad (5.9)$$

Here, $K_{ATP} = k_3/k_{-3}$ and $K_{ADP,P_i} = k_{-1}/k_1$ are the apparent constants of the product (ATP) and substrates (ADP and P_i) dissociation from the enzyme, E. An equilibrium constant, $K_{eq}^* = k_2/k_{-2}$, related to the interconversion stage $2(E \cdot ADP \cdot P_i \rightleftarrows E \cdot ATP)$ can be much greater than the equilibrium constant, K_{eq}, of the overall reaction. This could be realized if the ratio $K_{ATP}/K_{ADP,P_i} \ll 1$, which means that the dissociation of a product (ATP), is energetically less profitable than the substrates (ADP and P_i) dissociation. This circumstance implies that the ratio of *bound* ATP and *bound* ADP could be much greater than the ratio of ATP and ADP in aqueous solution, i.e., $[ATP]_b/[ADP]_b \gg [ATP]_s/[ADP]_s$. All experimental data available confirm that this is true.

According to numerous measurements of nucleotide binding to the coupling factor F_1 from different energy-transducing membranes [47, 150, 151], the ratio $[ATP]_b/[ADP]_b$ is about one unit. This point seems to be very important. If an equilibrium constant of the interconversion $K_{eq}^* \approx 1$, then the elementary step of ATP formation from ADP and P_i at the catalytic center of F_1 occurs *energetically gratis*. This means that the ATP formation from enzyme-bound ADP can proceed at the enzyme catalytic center without an immediate input of energy from the external source. Energy liberated in the course of the energy-donating process must be used for ATP release from the F_1 catalytic center. There is a line of experimental evidence for this method of ATP formation:

(i) the existence of the centers of very strong nucleotides binding to ATPsynthase;

(ii) releasing the tightly bound nucleotides induced by the membranes energization; and

(iii) finding that the P_i incorporation into ADP on the ATPsynthase (i.e., the formation of tightly bound ATP from ADP and P_i) does not need an input of energy.

Direct evidence for the last statement was obtained in a number of very important experiments carried out in Boyer's laboratory. Studying the P_i–$H_2^{18}O$ isotopic exchange reaction in uncoupled mitochondria, Boyer and his collaborators had demonstrated that, in the catalytic center of the mitochondrial ATPsynthase, there occurred numerous acts of the formation and rupture of the covalent bond between ADP and P_i. Boyer proposed that ATP formation in the catalytic center is simply the reversion of the ATP hydrolysis reaction, $ATP + H_2O \rightleftarrows ADP + P_i$. This conclusion follows from the experimental fact that ATPase catalyzes the incorporation of the ^{18}O isotope from H_2O into P_i that appears in the medium. The main features of Boyer's binding-change mechanism of ATP synthesis were clearly formulated in [167] as follows:

1. *Energy promotes release of* ATP *tightly bound to a catalytic site.*
2. *Energy promotes a binding of* P_i *and ADP in a manner that favors tightly bound* ATP *formation.*
3. *Interconversion of bound ADP and* P_i *to tightly bound* ATP *is largely or wholly independent on energy input.*
4. *The binding changes result from the energy-linked interconversion of a cooperative catalytic site.*

There are at least three groups of experimental findings that lie at the foundation of the binding-change mechanism of ATP synthesis.

(i) The study of medium P_i–$H^{18}OH$ and medium ATP–$H^{18}OH$ isotopic ^{18}O exchanges has demonstrated that ADP and P_i binding to F_1, and the release of ATP from F_1, did not occur without the energization of mitochondria or chloroplasts. Meanwhile, the dynamic interconversion of *bound* ADP and P_i into *bound* ATP ($E \cdot ADP \cdot P_i \rightleftarrows E \cdot ATP$) proceeded without the participation of $\Delta\bar{\mu}_{H^+}$, i.e., numerous acts of the formation and rupture of the covalent bond between ADP and P_i occured at the catalytic center of F_1 with the lack of energy input.

(ii) The initial kinetics of nucleotide releasing and binding to the chloroplasts coupling factor CF_1, registered immediately after the acid-base transition or in response to switching on the light, indicate that there are no other phosphorylated precursors but tightly bound ATP. Immediately after the chloroplast energization (i.e., with the lag phase less than 2.5 ms) there is a release of tightly bound ATP from a catalytic center that is accompanied with ADP binding from the medium.

(iii) The *de novo* formation of tightly bound ATP from ADP and P_i in de-energized membrane systems, or even in the isolated water-soluble coupling factor F_1 (we will consider these data in more detail in the next section).

To clarify the first point, let us again consider a scheme of the ATPsynthase reaction (5.8). Steps 1 and 2 in this sequence of processes (5.8) can produce an oxygen ^{18}O exchange between P_i and H_2O in the medium. Steps 2 and 3 produce a medium ATP–H_2O isotopic (^{18}O) exchange. The net chain of reactions leading to ATP synthesis or hydrolysis, that involves as the obligatory intermediate the reversible step 2 in the sequence of events (5.8), could cause the incorporation of more than one ^{18}O atom from H_2O into P_i. Evidence was obtained that reversible ATP cleavage is the major route of the ^{18}O atoms entry in the intermediate P_i–H_2O exchange. This reaction is resistant to the uncouplers of oxidative and photophosphorylation (e.g., dinitrophenol), and can be catalyzed even by the water-soluble coupling factors F_1 and the CF_1 enzymes [165, 166]. This means that numerous acts of reversible formation and rupture of the covalent bond between ADP and P_i occur at the catalytic center of F_1 without an energy input. Boyer concluded that energy input serves mainly to increase the forward rate constants, k_1 and k_3. This point of view is not compatible, however, with the reaction mechanism in which the protons pulled into an active center by $\Delta\bar{\mu}_{H^+}$ directly attack P_i, promoting the removal of an —OH group from P_i, and thus catalyzing ATP formation.

The stoichiometry of subunits in F_1 corresponding to an $\alpha_3\beta_3\gamma\delta\varepsilon$ structure, as well as the analysis of kinetic data obtained in various laboratories, prompted Boyer to develop the model for a three-site binding-change mechanism [167]. The sketch explaining this mechanism is depicted in Fig. 5.24. The model implies that three catalytic β subunits are in different conformational states during enzyme functioning. These three conformations are characterized by the following properties:

(i) loose binding ADP and P_i,
(ii) chemical equilibrium between tightly bound ATP, ADP, and P_i; and
(iii) loose binding ATP.

The interconversion from one conformation to another might be controlled, for example, by shifting the position of nucleotide binding α and β subunits relative to the γ, δ, and ε subunits in the core of the F_1 complex.

Thus, if the mechanism of ATP, synthesis proposed by Boyer is true, *an elementary step of* ATP *formation from* ADP *and* P_i *in the catalytic center of the coupling factor* F_1 *can proceed without imposing the* $\Delta\bar{\mu}_{H^+}$ *across the coupling membrane.* The foregoing does not of course exclude the important role of $\Delta\bar{\mu}_{H^+}$ in the processes of H$^+$ATPsynthases functioning in the membranes of energy-transducing organelles *in vivo*. As we will see below, the $\Delta\bar{\mu}_{H^+}$ generated across the coupling membrane provides conditions for the reiterative mode of ATPsynthase functioning, thus providing recurrent liberation of

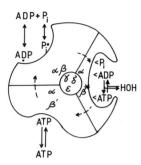

Fig. 5.24. A three-site binding-exchange mechanism for ATPsynthase (after [167]).

newly formed ATP molecules from F_1 into the solvent. The model explaining the mechanism of the cyclic operation of H^+ATPsynthase will be considered in more detail in Section 5.3.2.1.

The idea that energy delivered by energy-donating processes (electron transport) is used for releasing tightly bound ATP from an enzyme active center is now accepted by the majority of specialists in the field of bioenergetics. Energy released in the course of oxidative processes is transmitted to the F_1 active center by a process which does not include direct participation of the bulk phase-to-bulk phase gradient of protons, although the formation of an ATP molecule is associated with the translocation of three protons through the CF_0–CF_1 complex. Several possibilities were considered concerning the role of translocated protons in membrane phosphorylation [38, 45–47, 137, 153, 159, 161]. The mechanism of direct coupling proposed by Mitchell implies that translocated protons are delivered directly (through the membrane section of ATPsynthase, F_0) to the catalytic center at F_1, thus providing the conditions for ATP synthesis from ADP and P_i. On the other hand, various versions of the mechanism of indirect coupling suggest that the transmembrane (or intramembrane) motion of protons induces conformational changes in F_1 that facilitate performing the energy-requiring step in the link of processes of ATP synthesis, i.e., the liberation of tightly bound ATP.

The energy-dependent conformational changes in the coupling factor F_1 were considered by several authors as an important step in the mechanism of tightly bound ATP release from the catalytic center of F_1 [38, 54, 131, 163]. It has been proposed that the binding energy of substrates can be utilized for tightly bound ATP formation [168, 169]. Indeed, the energies of ionic bonds are rather large. For instance, the standard affinity of Mg^{2+} binding to the chloroplast coupling factor CF_1 is $\mathscr{A}^0 = 4.6$ kJ/mole; for two centers of P_i binding to CF_1, $\mathscr{A}_1^0 = 6.2$ and $\mathscr{A}_2^0 = 4.8$ kJ/mole, respectively [169]. There are several sources of evidence indicating the existence of energy-dependent conformational changes in the chloroplast coupling factor CF_1. Ryrie and Jagendorf found that chloroplast energization enhanced tritium exchange, stimulating incorporation of ^3H from ^3H_2O into CF_1 [170]. Using the fluorophore fluoresamine covalently bound to CF_1, Kraayenhof and Slater detected rapid light-induced conformational changes in membrane-bound CF_1 [171, 172]. Similar results were obtained by Wagner and Junge who used fluorescent labeling to study the conformational changes in CF_1 [173–175]. Using optical methods (circular dichroism and UV absorption spectroscopy), Kagawa and his collaborators demonstrated that ADP and ATP binding to an isolated catalytic β subunit of F_1 induced the conformational changes of the protein globule [176]. Later, Gantchev and his collaborators [177], using the spin-labeling technique, were able to demonstrate that nucleotide binding caused the conformational changes in the isolated chloroplast coupling factor CF_1. Some models which consider the conformational changes in F_1 caused by the protonation/deprotonation processes will be considered in Sections 5.3.1.4 and 5.3.1.5.

5.3.1.3. ATP Synthesis from ADP and P_i Catalyzed by Water-Soluble Coupling Factor F_1

According to earlier variant(s) of the hemiosmotic hypothesis [45, 46], the protons pulled into the active center of the coupling factor F_1 by the transmembrane difference $\Delta\bar{\mu}_{H^+}$ directly attack the O^- group of the inorganic phosphate, P_i, promoting P_i interaction with ADP. This notion might imply that protons play the role of one of the substrates in the reaction of ATP formation, ADP + P_i + $nH^+ \rightleftarrows$ ATP + H_2O. If the role of the protonic gradient is simply to concentrate the protons at the active center of F_1, we could also expect that the synthesis of ATP molecules might occur without the transmembrane $\Delta\bar{\mu}_{H^+}$ gradient. Moreover, even an isolated water-soluble coupling factor might catalyze the formation of ATP from ADP and P_i, providing the appropriate H^+ concentration was adjusted in the surrounding medium. Indeed, there is certain evidence that an elementary act of F_1-catalyzed ATP formation can proceed in the absence of the proton electrochemical gradient.

In 1982 Feldman and Sigman discovered that the purified water-soluble coupling factor CF_1 from spinach chloroplasts could catalyze the formation of ATP from ADP and P_i [178]. It was demonstrated that $^{32}P_i$ from the surrounding solution formed the covalent bond with ADP leading to the synthesis of the enzyme-bound ATP[^{32}P]. Since the reaction yield was independent of the ADP concentration in the solution, as well as being insensitive to the addition of hexokinase and glucose, the authors concluded that ATP[^{32}P] was formed from enzyme-bound ADP. The maximal yield (parameter α) of the reaction had been observed when the enzyme was incubated in the medium at pH 6, and the reaction yield, α, was about 0.24 molecules of ATP[^{32}P] per one molecule of CF_1.

The formation of bound ATP[^{32}P] from ADP and $^{32}P_i$ can also be catalyzed in the presence of dimethylsulfoxide by the coupling factors F_1 purified from mitochondria [179] and the thermophylic bacteria PS3 [180]; under optimal conditions, the maximal yields of F_1-bound ATP were $\alpha \cong 0.6$ and $\alpha \cong 0.8$, respectively. It has been demonstrated in [180] that the formation of enzyme-bound ATP$[^{32}P]$ in the catalytic center of F_1 from the thermophylic bacteria PS3 is accompanied by a concomitant decrease in the number of enzyme-bound ADP. A tentative reaction mechanism for the formation of a tight F_1–ATP complex, proposed by Sakamoto and Tonomura in [179], can be written as follows:

$$E + ADP + P_i \underset{\substack{\text{rapid} \\ \text{equilibrium}}}{\xrightarrow{\hspace{1.5cm}}} E_{P_i}^{ADP} \xrightarrow{\hspace{1.5cm}} E_{P_i}^{<ADP} \underset{\substack{\text{rapid} \\ \text{equilibrium}}}{\xrightarrow{\hspace{1.5cm}}} E^{<ATP}.$$

Here, $<$ADP and $<$ATP denote tightly bound nucleotides. It should be noted that in all these experiments the ATP$[^{32}P]$ formation was proved not to be associated with the adenilate–kinase-like reaction (2ADP $\rightleftarrows$ ATP + AMP). This adenilatekinase-like reaction catalyzed by the chloroplast coupling factor CF_1 was demonstrated in the laboratory of Moudrianakis [181].

Thus, the experiments with purified enzymes from different sources demonstrate that *de novo* formation of *bound* ATP *from* ADP *and* P_i *can be catalyzed by the water-soluble coupling factor* F_1. If this mechanism is realized in the reaction center of membrane-bound ATPsynthase, *an elementary step of* ATP *formation from* ADP *and* P_i *in the catalytic center of the coupling factor* F_1 *could proceed without imposing* $\Delta\bar{\mu}_{H^+}$ *across the coupling membrane.*

The foregoing does not, of course, exclude the important role of $\Delta\bar{\mu}_{H^+}$ in the processes of H$^+$ATPsynthases functioning in the membranes of energy-transducing organelles *in vivo*. As we will see below, the $\Delta\bar{\mu}_{H^+}$ applied across the coupling membrane provides the proper conditions for organizing the repetitive binding substrates to F_1 and the liberation of newly formed ATP molecules from the F_1 catalytic center to the solvent, thus the ensuring mechanism of cyclic operation of H$^+$ATPsynthase.

5.3.1.4. ATP Synthesis Induced by the Acid-Base Transitions

In chloroplasts, the coupling factor CF_1 is exposed to the external medium with pH ≥ 8 (pH$_{out} \cong 8.0$–8.5 is optimal for ATP synthesis). These values of pH are unfavorable for the formation of ATP from ADP and P_i bound to F_1: the maximal yield of the tightly bound ATP$[^{32}P]$ formation by the isolated chloroplast coupling factor CF_1 was observed for the enzyme incubated at pH 6 [178]. However, the data in [178–180] do not exclude, generally speaking, the direct participation of protons in the process of ATP formation in the active center of the coupling factor F_1. The acidification of the thylakoid's interior (lumen or intramembrane domains) might promote the formation of ATP in the catalytic center of CF_1, simply by delivering the protons to the CF_1 active center through the membrane portion of ATPsynthase, CF_0. Being "pushed" to the CF_1 active center (or its vicinity) by high internal protonic pressure, these protons could provide the conditions for the energy gratis formation of tightly bound ATP from tightly bound ADP and P_i. On the other hand, the "alkaline" pH values in the external surroundings of CF_1 would facilitate the liberation of tightly bound ATP from the catalytic center. This view finds direct support from the results of numerous experiments on ATP synthesis by the generation of an artificial pH gradient as the result of acid-base transition.

Membrane-Bound H^+ATPsynthases

Let us consider the results of the experiments carried out on coupled and uncoupled mitochondria [182]. The main results are as follows: Intact mito-chondria are able to perform quantitative ATP synthesis in response to a fast pH increase. From two to five (depending on preparation) newly synthesized ATP molecules were recorded per pH jump per one ETC. The obligatory condition was that the value of pH $\cong 8.2$ should be exceeded in the course of the pH jump. The same results were obtained for mitochondria uncoupled by aging, multiple freezing and thawing, or by addition of the uncouplers (FCCP or gramicidin A), unable to perform oxidative phosphorylation. Fig-ure 5.25(A) demonstrates the experimental dependencies of the ATP yield (parameter α) induced by a fast pH increase in the suspension of uncoupled rat liver mitochondria on the final pH value (at fixed initial one, curve 1), and on the initial pH value (at fixed final one). These dependencies obviously correspond to the titration curves of acid groups with pK $\cong 8.2$. The ATP formation induced by the pH increase does not require the functioning of ETC because it was not inhibited by cyanide or antimycin A. ATP synthesis was completely blocked by ATPsynthase inhibitors (oligomycin; DCCD, the inhibitor of the H^+ channel, and arsenate, the concurrent on P_i for binding

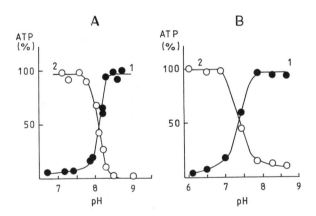

Fig. 5.25. Dependence of the ATP yield on the final pH value at fixed initial one (1), and on the initial pH value at fixed final one (2) as a percent of the maximum yield: (A) uncoupled rat liver mitochondria; and (B) membrane-wall fragments from *Staphylococcus aureus*.

to F_1) and could not be observed in the absence of any one of the phosphorylation substrates ADP and P_i. With the surplus of phosphorylation substrates it is possible to obtain, in this way, scores of ATP molecules per ETC with one and the same sample by changing pH repeatedly, for example, from 7.5 to 8.5 and back again. There is no ATP formation with the opposite sign of the pH jump (i.e., after a fast pH decrease).

If ADP and P_i are added after a "proper" pH jump, the ATP yield depends on the time interval between the "pH stroke" and the addition of phosphorylation substrates. For mitochondria the relaxation time, $\tau_{1/2}$, i.e., the interval after which the ATP yield decreases by half, is about 10 s at room temperature. The existence of this rather long "memory" may be explained by the complex structure of mitochondria. Isolated ATPases display faster (by one order of magnitude) relaxation after a pH-jump.

The ATP formation in response to a fast pH increase was detected in membrane-bound H^+ATPases obtained from different biomembranes: membrane-wall fragments from *Staphylococcus aureus* (Fig. 25(B)) which did not contain endogenous substrates and were permeable for ions [183], membrane fragments from *Micrococcus lisodecticus* [184], and membrane fragments from bacteria *Lactobacillus casei* which possess no ETCs [185]. For all these preparations, a fast acid-base pH-jump in the suspension of membrane fragments leads to the ATP synthesis, while for the pH-jumps of the opposite sign (a pH decrease) there was no ATP formation.

Isolated ATPases

Topologically closed membrane vesicles and even membrane fragments are not necessary to perform the elementary acts of ATP synthesis initiated by the fast pH increase. Results similar to those described above were obtained with isolated water-soluble components of the ATPases from *Micrococcus lisodecticus* [184], *Lactobacillus casei* [185] and the coupling factors $CF_1 - CF_0$ and CF_1 from *bean chloroplasts* [186]. The phenomenon of ATP synthesis caused by the pH jump seems to be a common property not only of ATPases from energy-transducing membranes, but also of other ATPases whose normal functioning is not associated with the ATP synthesis: *myosin* [187], Ca^{2+}ATPase of *sarcoplasmic reticullum* [188–189], and *nitrogenase* [82].

5.3.1.5. ATP Synthesis from ADP and P_i as Considered from the Viewpoint of the Relaxation Concept of Enzyme Catalysis

The results of experiments on acid-base phosphorylation (the pH jump-induced ATP synthesis) performed, by isolated or membrane-bound H^+ATPases could be interpreted on the basis of the relaxation concept of enzyme catalysis. This concept allows us to suggest a model for ATP synthesis consistent with all the experimental data considered above. It has been proposed, in [182–187, 190, 191], that the fast ionization of certain acid group(s) of the F_1 component of H^+ATPsynthase leads to the enzyme transition to a nonequilibrium (macroergic) state. This suggestion has been prompted by experiments with ferricytochrome C [192], which revealed that the fast deprotonation of certain groups on ferricytochrome C led to the creation of a conformationally nonequilibrium state of the protein globule and its subsequent relatively slow relaxation to equilibrium. An energy-accepting stage of ATP expulsion from the catalytic center of the coupling factor F_1 is realized during relatively slow relaxation of this state to equilibrium. Formally, the reaction mechanism can be written in the following way:

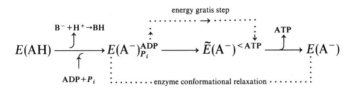

The very fact that the ATP formation occurs only in response to a *fast* enough pH jump indicates the involvement of several acidic groups that must dissociate simultaneously to provide the ATP liberation from F_1. The necessity of reducing agents to realize the ATP synthesis in response to a pH jump indicates that certain redox groups in F_1 should be in the reduced state. If they are oxidized, the acid-base pH jump cannot lead to the ATP dissociation from F_1. Bearing in mind the role of thiol groups in the modulation of ATPsynthase activity (see, for references, [193]), there are good reasons to believe that it is the preliminary reduction of the enzyme —S—S— bridges accompanied by the formation of —SH acid groups that might be associated with the occurrence of the pH jump-induced macroerg. Really, only by being reduced can these groups be involved in the protonation/deprotonation processes ($-SH \rightleftarrows -S^- + H^+$).

In the spirit of the conformational relaxation concept the "primary macroerg" is identified with the ATPsynthase molecule in a nonequilibrium state. There are obvious physical reasons for the appearance of the "macroergic" nonequilibrium state of the enzyme molecule that arises immediately after the proton(s) dissociation from F_1: the potential energy of the system (enzyme–substrate complex) increases with the appearance of extra negative charge(s) in the moiety of a protein globule with a practically unchanged conformation. This ensures performance of the energy-accepting stage of the ATP synthesis (the liberation of tightly bound ATP from F_1). Energetic aspects of the problem will be analyzed in Section 5.3.2.2.

Let us now consider a feasible sequence of events that could take place during one cycle of the enzyme turnover. There are good reasons to believe that exposing an active center of F_1 to the medium with the appropriate activity of hydrogen ions (e.g., by the preliminary incubation of an isolated enzyme in the "acidic" solvent, or by an active center "communication" with the internal bulk phase of the intrathylakoid volume through the membrane segment of the ATPsynthase complex, etc.) promotes the binding of the substrates of phosphorylation (ADP and P_i) to F_1. According to [194], the apparent constant of ADP binding to the chloroplast coupling factor CF_1 increases with a lowering of pH from 8 to 6.5. As has been demonstrated by Feldman and Sigman [178], the slightly "acidic" pH of the medium (pH 6) is optimal for the formation of tightly bound ATP by the purified water-soluble coupling factor CF_1. We assume that there is (are) certain proton-accepting group(s) in CF_1 whose protonation/deprotonation status determines the substrates while are binding. Being protonated, this group(s) facilitates attaching substrates to the F_1 catalytic center. By the way, the same (or similar) protonated group(s) (denoted, certainly, as BH) could be directly involved in the reaction of the ATP formation taking place at the catalytic center

$$ADP + P_i + kBH \rightleftarrows ATP + H_2O + kB^-. \tag{5.10}$$

The stoichiometry coefficient k should be determined by the relationship between the apparent pK values of ATP, ADP, and P_i located at the catalytic center. We can say that the protonated group(s) BH serves as the role of the immediate H^+ donor in the reaction (5.10). As we have discussed above, the reaction of the ATP formation at the catalytic center itself does not need the energy input. Appropriate surroundings of ADP and P_i at the catalytic center probably provide the electrostatic screening of negative charges of the phosphate groups. This helps to overcome the repulsion between negatively charged phosphate groups, thus facilitating energy gratis formation of the covalent bond between ADP and P_i.

The conditions for the tightly bound ATP release from F_1 (energy-requiring stage in the chain of events of the ATPsynthase reaction) could be created after proton dissociation from another (other) ionizable group(s), AH, located in the vicinity of the active center, $AH \rightarrow A^- + H^+$. This process can be initiated as the result of the AH contact (direct or indirect) with an "alkaline" medium that surrounds F_1 (e.g., the aqueous solution of the chloroplast stroma, or the mitochondrial matrix). The neighborhood of two negative charges, A^- and B^-, is equivalent to increasing the potential energy of the system (enzyme + newly formed tightly bound ATP). This state of enzyme, $\tilde{E}[A^-(ATP*H_2O)B^-]$, occurring immediately after the AH ionization, could be identified with a primary macroerg. In the course of $\tilde{E}[A^-(ATP*H_2O)B^-]$ relaxation to the equilibrium tightly bound ATP dissociates from an enzyme molecule, $\tilde{E}[A^-(ATP*H_2O)B^-] \rightarrow E[A^-, B^-] +$ ATP. Here, $E[A^-, B^-]$ symbolizes an equilibrium state of an enzyme after the completion of the conformational relaxation process.

We want to stress that there might be another feasible sequence of AH and BH deprotonation events: AH deprotonation can precede the step of substrate binding. This way of creating a nonequilibrium macroergic state $\tilde{E}[A^-(ATP*H_2O)B^-]$ could be realized if a proton(s) dissociates relatively slowly from the BH group(s) which determines substrate binding. In this case, another acidic group(s), BH, could still be protonated for a certain period of time after the H^+ dissociation from AH. If ADP and/or P_i have time to bind to F_1 before the BH deprotonation, an energized state $\tilde{E}[A^-(ATP*H_2O)B^-]$ would appear immediately after realizing the act of ATP formation (5.10). Such a "fruitful" sequence of events could easily explain why the step of deprotonation can, in principle, precede the step of substrate binding.

On the other hand, the appearance of a nonequilibrium high-energy state $\tilde{E}[A^-, B^-]$, "exited" as the result of AH and BH deprotonation from the substrate-bare enzyme molecule $E[AH, BH] \rightarrow \tilde{E}[A^-, B^-] + 2H^+_{out}$, cannot be used to perform useful work. The potential energy of an enzyme molecule in state $\tilde{E}[A^-, B^-]$ will dissipate during the course of the following enzyme relaxation to equilibrium, $\tilde{E}[A^-, B^-] \rightarrow E[A^-, B^-]$. This process may include conformational changes in the protein globule itself (e.g., the displacement of charged groups, their exposure to a polar solvent, etc.), as well in as the interaction of the protein globule with its surroundings (binding counterions, etc.). For instance, an apparent constant for Mg^{2+}, binding tightly to the chloroplast coupling factor CF_1, increases as the pH changes from 6.5 to 8 [194]. Thus, after CF_1 deprotonation (e.g., as the result of a pH jump) there would take place an additional binding of Mg^{2+} to CF_1 that will compensate for the appearance of a negative charge on the enzyme molecule. We want to note that the "useful" and "futile" trajectories of enzyme relaxation to the equilibrium state $E[A^-, B^-]$ will differ for the "loaded," $\tilde{E}[A^-(ATP*H_2O), B^-]$, and "unloaded," $\tilde{E}[A^-, B^-]$, enzyme molecules.

According to the suggestion put forward by Kozlov [195], electrostatic interactions play a crucial role in nucleotide binding–releasing events in the catalytic center of F_1. The imbalance in electric charges, occurring due to the difference in the total charges of the substrates (ADP, P_i and Mg^{2+}) versus the products (ATP, Mg^{2+}), that facilitate the liberation of the newly synthesized ATP molecule from CF_1, might also explain why the application of an electric field (or the appearance of local electric charges) or a pH-jump promotes releasing the product of ATPsynthase reaction, i.e., tightly bound ATP. The ionization of acidic groups is possible, in principle, not only due to increasing the pH of the enzyme surrounding medium. The same effect of deprotonation can be achieved as a result of decreasing the pK value of acid groups at fixed pH. Such an alteration of the pK value could be realized, for instance, by imposing an electric field across the coupling membrane containing ATPsynthase (see the scheme in Fig. 5.26).

There is a certain similarity between the above-described model and the model suggested by Kagawa in 1984 [47, 131]. His model explained the release of ATP from the F_1 catalytic center by conformational changes induced by the proton transfer events occurring in the so-called acid-base cluster located near the ATP binding site. In order to satisfy the required stoichiometry of ATP synthesis, $3H^+$/ATP, at least three proton-accepting groups should cooperate with one coupling device. Being clustered in a narrow area of F_1 in the vicinity of the reaction center, these groups are able to receive

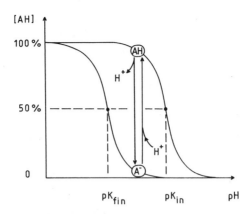

Fig. 5.26. Scheme illustrating the potential-driven acidic group(s) AH deprotonation at fixed pH: potential-induced decrease in the pK value of AH group(s) from pK_{in} to pK_{fin} leads to AH deprotonation, $AH \rightarrow A^- + H^+$, at fixed pH in the aqueous bulk phase surrounding H^+ATPsynthase complex.

protons from the membrane section of the ATPsynthase, the F_0 complex. Kagawa relates these clusters to homologous sequences in the catalytic β subunits of the coupling factor F_1 from mitochondria, chloroplasts, and E. coli. According to [47, 131], the acidic cluster is composed of $Asp-Glu-Leu-Ser-Glu-Glu-Asp$ residues and the basic cluster of $Arg-Ala-Arg-Lys-Ile-X-Arg$ residues located very close to the Rossmann fold ($\alpha-\beta-\alpha-\beta-\alpha-\beta$ structure) in the catalytic β subunit of F_1.

It has been suggested by Kagawa that an application of $\Delta\bar{\mu}_{H^+}$ across the coupling membrane induces the transfer of three protons from $-NH_3^+$ to $-COO^-$ groups and the formation of neutral residues ($-NH_2$ and $-COOH$). The neutralization of charged acidic and basic groups causes the conformational change in the Rossmann fold of F_1 that facilitate releasing the tightly bound ATP from the catalytic center. Indeed, the conformational changes in F_1 associated with ATP or ADP binding to the β subunit were detected in several works [170–177]. Protonation of $-NH_2$ residues from the F_0 side, and proton dissociation from $-COOH$ to the F_1 bulk phase would provide the reiterative regime of the ATPsynthase operation. The requirements of the acid-base cluster model are consistent with numerous experimental facts concerning the structure and physical–chemical properties of the coupling factor F_1, as well as with the experimental study of the chemical modificators' influence on the ATPsynthase activity [47].

It is important to note, however, that there is an essential difference between the acid-base cluster model and the conformational relaxation model. Kagawa considers the acid-base cluster as a device that can be driven by H^+ concentration changes according to the mass action law. However, the overall reaction of ATP formation cannot be explained simply by the mass action law: the ATP appearance in aqueous solution can be initiated by decreasing the concentration of H^+ ions in the external medium, e.g., as a result of the acid-base shift in Jagendorf's experiment. According to the reaction stoichiometry, decreasing the H^+ concentration can shift the equilibrium only toward the ATP hydrolysis. While the proton mass action might facilitate the tightly bound ATP formation (according to [178–180], the synthesis of ATP$[^{32}P]$ was optimal at pH $\approx$ 6–7), the liberation of newly formed ATP from F_1 can proceed only with increasing pH in the surrounding medium, $pH_o > pK_a \cong 7$–8. Of course, releasing tightly bound ATP from the chloroplast coupling factor CF_1 could occur after very significant acidification of the reaction medium [196]. However, speaking of a pH jump-induced ATP liberation we consider only those processes that take place in the physiological range of pH values. These processes are initiated not due to the acidification but to the formation of an extra negative anion and further conformational relaxation.

5.3.2. ATP Synthesis under Steady State Conditions

The reaction mechanism considered above describes only the ATP synthesis initiated by the single act of a fast and synchronous deprotonation of certain acid groups in the F_1 component of H^+ATPases. This process was designated as the "elementary act" of membrane phosphorylation [190, 191]. We consider below a feasible mechanism of ATPsynthase cyclic functioning with the reiteration of the "elementary acts" of ATP formation. This mechanism must describe the normal processes of oxidative and photosynthetic phosphorylation, as well as the repeated acts of ATPsynthase actuation in the course of post-illumination ATP synthesis in chloroplasts, or a multitude of acts of ATP formation performed by each active ATPsynthase complex after the acid-base transition (chloroplasts, liposomes which contain H^+ATPases, and bacteriorhodopsin, etc.) [19–35].

5.3.2.1. The Possible Model for ATPsynthase Cyclic Functioning

Let the mechanisms of the elementary act of ATP synthesis in oxidative and photophosphorylation, as well as in the experiments on post-illumination or acid-base-induced ATP synthesis, be the same as those described above in experiments with isolated ATPases. Really, the processes of ATP synthesis in all these cases are sensitive to the same inhibitors of ATPase. In order to ensure the fulfillment of the reiterative mode of operating the coupling factor, ATPsynthase should contain a special controlling "device," which would provide triggering of the reiterative protonation–deprotonation events in the acid-base cluster of F_1. Since multiple ATP synthesis initiated by an artificial H^+ gradient can be realized without ETC functioning, to provide the periodicity of enzyme functioning we have to assume the controlling "device" must be an *intrinsic part* of the ATPsynthase. The data available indicate that such

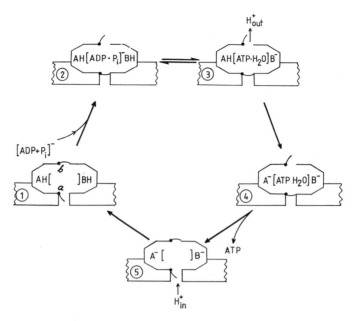

Fig. 5.27. Scheme of H⁺ATPsynthase cyclic functioning.

a "device" should be actuated by the processes of binding and releasing the adenine nucleotides from the active center of the coupling factor F_1.

Let us now consider a simplified scheme (Fig. 5.27) of ATPsynthase cyclic functioning which explains the multiple ATP formation after creating a transmembrane pH difference in the vesicles (thylakoids, submitochondial particles, chromatophores, or proteoliposomes) containing H⁺ATPases with F_1 factors exposed to the external aqueous medium. For the sake of simplicity, in the scheme shown in Fig. 5.27 we consider only one nucleotide-binding center and two proton-accepting groups, A⁻ and B⁻. As a "switch" for exposing these functional groups alternately to the internal ("acidic") or external ("alkaline") aqueous bulk phase we can consider the attachment of ADP to F_1 and the ATP dissociation from its catalytic center. Different positions of the switch symbolize the abilities of the protonizable groups A and B to contact the internal and external media. This is shown in Fig. 5.27 as two "gates" designated by the symbols "*a*" and "*b*." Gate "*a*" provides the contact of the functional acid groups with the interior of the vesicle, while gate "*b*" provides the contact with the external medium. The open position of "*a*" with the closed "*b*" key (state 1) indicates that the average rates of A⁻ and B⁻ protonation from the internal proton pool agrees with the corresponding rates of proton dissociation from AH and BH to the external surroundings. The closing of "*a*" can be caused by the attachment of nucleotides, while their liberation from F_1 would provide the contact of A⁻ and B⁻ with the vesicle interior symbolized as open "a." Actually, it is well known that under any conditions unfavorable for ATP synthesis (e.g., in the absence of P_i) the adenine nucleotide (ATP or ADP) binding with the chloroplast CF_1 active center blocks the efflux of protons from the thylakoid lumen through the CF_1–CF_0 complex [197–201].

Let us now consider a feasible sequence of events during the cyclic functioning of membrane-bound H^+ATPsynthase in the presence of a transmembrane pH difference. Position 1 in Fig. 5.27 corresponds to the enzyme quasi-equilibrium state: there are no phosphorylation substrates in the active center, the functional acid groups are protonated due to their contact with the "acidic" interior of a vesicle ($pH_i < pK_a$), and "a" is open; a fast leakage of protons into the external aqueous phase via ATPsynthase is prevented by the barrier hindering the contact of the AH group with the exterior, symbolized in Fig. 5.27 by key "b" in the locked position. The attachment of phosphorylation substrates to the active center of the coupling factor (transition $1 \rightarrow 2$) makes possible two events:

(i) the energy gratis interconversion of bound ADP and P_i into tightly bound ATP (reaction (5.10), transition $2 \rightarrow 3$); and

(ii) the blockade of the proton pathway "a," that is symbolized as the "turn over" of a protonic switch.

The latter simply implies that bound nucleotides break down the contact of acidic groups with the interior volume of the vesicle ("a" is locked). Thus, after the substrates bind, the groups AH eventually become ionized due to the efflux of protons into the "alkaline" outer aqueous phase ($pK_a < pH_o$), i.e., "b" is open (states 2–4). In accordance with well-known experimental fact [197–201], that the process of ADP phosphorylation accelerates the efflux of protons via the CF_1–CF_0 complex, the model suggests that the substrates binding stimulates the liberation of protons into the external medium under phosphorylating conditions.

The assumption that the protonic "switch" is controlled by the nucleotide binding is supported by the data obtained in chloroplasts and the isolated coupling factor CF_1. Nucleotides binding with an active center induce conformational changes in CF_1 which involve a large part of the protein globule [170–177]. These changes can probably stimulate the dissociation of protons from AH to the external "alkaline" medium. Being "isolated" from the "acidic" interior by bound nucleotides, this group(s) will easily dissociate, thus leading to the occurrence of the enzyme nonequilibrium state (Fig. 5.27, state 4). In this state, the enzyme contains a newly formed, tightly bound ATP molecule. State 4 is energy unfavorable due to excessive negative charges in the protein globule. In the course of the following relaxation of the enzyme complex, the tightly bound ATP molecule dissociates from F_1 to the external aqueous phase. In this way the enzyme reaches state 5: "a" is open and "b" is closed. After a subsequent protonation of the acid group (transition $5 \rightarrow 1$) the cycle can be repeated as long as condition $pH_i < pK_a < pH_o$ is fulfilled.

It is easy to extend this model in order to interpret any kind of experimental data on multiple reiterative ATP synthesis initiated by the creation of an artificial transmembrane difference of the electrical potentials, $\Delta\varphi$, generated across the closed vesicles, e.g., liposomes encrusted with membrane-embedded $H^+ATPases$. In this case, when the reiterative acts of ATP formation are driven by $\Delta\varphi$ in the lack of ΔpH, the protonation of the ionized acid group A^- is evidently provided by the field-induced decrease in the pK_a value of the ionizable group AH (Fig. 5.26). The $\Delta\varphi$ difference imposed across the coupling membrane can also lead to the occurrence of the difference in the proton concentrations related to the local regions in the vicinities of gates "a" and "b." This difference, induced by the electric field, can exist even without any pH difference between the bulk phases on both sides of the coupling membrane.

It is quite clear that for mitochondria, where F_1 is localized within the vesicle, a jump-like pH increase can lead to only one "elementary act," that indeed was observed in [182]. In this case the F_1 component of the ATP-synthase is faced inside the vesicle and, thus, according to Fig. 5.27, the state 5 → state 1 transition becomes impossible. Only a single act of the AH ionization would take place after the pH jump due to a passive leakage of protons via the mitochondrial membrane. Otherwise, for submitochondrial particles turned inside-out [21] a multiple ATP synthesis after the pH increase is achieved by means of the reiteration of the proton cycle. According to the model considered above, the transmembrane electrochemical gradient of protons plays the role of an extensive but not intensive factor [190, 191]. Its increase (or decrease) changes the number of elementary acts of ATP synthesis which can be performed, until the condition formulated above for the "elementary act" ceases to be fulfilled.

5.3.2.2. Photophosphorylation in Chloroplasts and Oxidative Phosphorylation in Mitochondria

The formation of one ATP molecule in chloroplasts and mitochondria is coupled with the translocation of three protons via ATPsynthase [11–15]. There are two extreme cases that can be achieved under physiological conditions in a steady state.

(i) Low concentrations of phosphorylation substrates (ADP or P_i) which limit the net rate of ATP formation. This corresponds, according to Chance and Williams [37], to the so-called state 4 characterized by a relatively low rate of electron transport (states of photosynthetic control in chloroplasts and respiratory control on mitochondria). In state 4 the net efflux of protons from the thylakoids (or their influx into the mitochondrial matrix) through functioning ATPsynthase is also low. By the way, this means that the energy-accepting and energy-donating processes are tightly coupled in both directions, although the pathways of forward and backward reactions can be essentially different [202]. Due to a relatively low loss of protons, in state 4 the $\Delta\bar{\mu}_{H^+}$ created can be maintained at a rather high level (about 180–240 mV), and thus ensures ATP synthesis by the mechanism discussed above.

(ii) In the presence of the surplus concentrations of ADP and P_i (state 3) the rates of electron transport and ATP synthesis are high. In this state, due to a high rate of coupled proton flux via ATPsynthase, the $\Delta\bar{\mu}_{H^+}$ under certain conditions can drop to a rather low value (see Section 5.2).

Thus, the question, arises: Is the transmembrane difference in the electro-chemical potential between the bulk phases, $\Delta\bar{\mu}_{H^+}$, always supported at a high enough level in steady state 3 in order to play the role of the immediate source of energy for the energy-accepting steps of ATP synthesis, or are the conditions $pH_i < pK_a < pH_o$ fulfilled to provide the cyclic operation of ATPsynthase upon phosphorylating conditions?

As we discussed in Section 5.2, there is experimental evidence demonstrating that under certain conditions an efficient steady state ATP synthesis can proceed upon the transmembrane $\Delta\bar{\mu}_{H^+}$ difference between water phases lower than the threshold level that is assumed, according to the hemiosmotic concept of membrane phosphorylation, to be necessarily held for stoichiometric phosphorylation. All these results were usually interpreted as the indication of the existence of an alternative (nonhemiosmotic) pathway of proton transport coupled to ATP synthesis (for details, see Section 5.2.5). We can give a simple explanation of these results within the frame of the above-described model for the reiterative mode of H^+ATPsynthase functioning.

The changes in the intrathylakoid pH_i may be important for photosynthetic control [71–77, 190]. The essential decrease in the pH_{in} value, that occurs in state 4 during noncyclic electron transport in chloroplasts ($\Delta pH \cong$ 2.5–2.9), leads to retardation of the electron flow rate. In state 3 the rather small decrease in the intrathylakoid pH_i is not sufficient to slow down the rate of electron transport between two photosystems, but ensures the conditions $pH_i < pK_a < pH_o$ necessary for providing the cyclic operation of ATPsynthase. It follows from Fig. 5.28 that at "external" $pH_o = 7.5$–8.5 (ensuring the high rates of ATP synthesis) the values of "internal" pH_i are below 7.2–7.5, i.e., less than the apparent pK_a value of ATPsynthase acid groups which might be involved in attaining the elementary act of ATP formation. All these results corroborate the idea that topologically closed membrane vesicles are necessary for the enzyme to return to the initial state with protonated acidic groups in an acid-base cluster of F_1. This makes possible stoichiometric ATP synthesis under steady-state or quasi-stationary conditions [190, 191].

On the Nature of the "Threshold" Response of H^+*ATPsynthases.* We have noted above that closed membrane vesicles are necessary to provide the cyclic functioning of ATPsynthases under steady state conditions, and that the transmembrane difference in proton electrochemical potentials plays the role of an *extensive rather than intensive factor of* ATP *synthesis.* According to the data discussed in Section 5.2, in chloroplasts the value of $\Delta pH = 1.0$

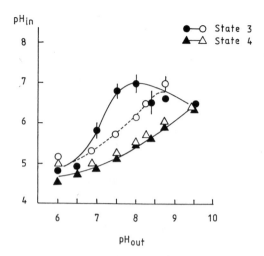

Fig. 5.28. The intrathylakoid pH values, pH_{in}, versus external pH in the suspension of bean chloroplasts, noncyclic electron transport ($H_2O \rightarrow$ methylviologen). The dark symbols are the data obtained with the kinetic method (after [76, 77]), and the light symbols are the data obtained with the use of the TEMPOamine partition technique (after [78]).

(i.e., equivalent to $\Delta\varphi = 60$ mV) appears to be quite sufficient to provide stoichiometric steady state ATP synthesis. On the other hand, many of the experiments reveal the existence of a so-called "threshold" $\Delta\bar{\mu}_{H^+} \cong 180\text{--}200$ mV, below which the ATP yield is, as a rule, negligible (see, e.g., [30–35]).

The question now arises: How could we come to agreement between such different results, and what is the nature of a threshold response to ATP formation? Either the appearance of a threshold can be interpreted as an indication of the response of ATPsynthase itself, or it may simply be the consequence of the exponential relation between the turnover number of ATPsynthase and $\Delta\bar{\mu}_{H^+}$ across the coupling membrane. According to chemiosmotic theory, the threshold should appear for purely energy reasons: the threshold is interpreted as the minimal value of free energy changes that provide for ATP synthesis due to the $\Delta\bar{\mu}_{H^+}$ driven passage of two to three protons via ATPsynthase. The transmembrane gradient of electrochemical potentials must exceed the threshold level determined by the phosphate potential, $\mathscr{A}_P$, i.e., $m\Delta\bar{\mu}_{H^+} \geq -\mathscr{A}_P$, where $\mathscr{A}_P = \mathscr{A}_P^0 + (RT/F) * \ln\{([ADP] * [P_i])/[ATP]\}$ and $\mathscr{A}_P^0$ is a standard affinity of the ATP synthesis reaction. In other words, ATPsynthase could operate only if protons, being forced by a sufficiently high transmembrane electrochemical proton gradient, pass through the CF_0CF_1 complex overcoming a certain potential barrier. ATPsynthase activity would be lacking below this threshold.

Discussing the relevant problems of chemiosmotic theory, Mitchell [153] asked the fundamental question: "What pushes what?" Since the substrates of phosphorylation could participate in the ATPsynthase reaction, being in various possible protonation states, Mitchell proposed that the osmotic protonmotive force may also be applied to the nucleotides and P_i, thereby pulling them into, or pushing them out from, the catalytic center of F_1. Thus, according to [153], the electrochemical proton gradient was considered as performing "the chemical work needed to pull the oxide ion out of P_i, and the physical work needed to pull P_i and ADP into the catalytic domain and push ATP out of it."

On the other hand, as we discussed in Section 5.2, a "localized" mechanism of ATP synthesis does not imply an obligatory energetic role of the transmembrane proton gradient. In this case, the turnover number of ATPsynthase would be determined solely by the flux of protons through the CF_0CF_1 complex, J_{H^+}, There is explicit evidence of a linear relationship between the rate of phosphorylation, J_{ATP}, and proton flux through the coupling membrane [203]. If there is a linear relationship between J_{ATP} and J_{H^+}, and J_{H^+} exponentially depends on $\Delta\bar{\mu}_{H^+}$, we must phenomenologically obtain an exponential relationship between J_{ATP} and J_{H^+} regardless of the coupling mechanism. This circumstance seems to be the reason for a threshold-like dependence of the turnover number of the ATPsynthases and proton potential across the coupling membrane [190].

There are also other possible reasons for the "threshold" response of ATPsynthase. One of them is the possible infringement of the condition $pH_i <$ $pK_a < pH_o$. For instance, having external $pH_o > pK_a$, at rather small values of ΔpH, the obligatory condition $pH_i < pK_a$ might not be fulfilled. The second reason can be associated with the regulatory effect of ΔpH on the ATPsynthase activity: an increase in ΔpH might increase the number of active ATPsynthase complexes [24, 32]. Finally, in experiments on acid-base-induced phosphorylation, the "threshold" may be simply a formal manifestation of the logarithmic relationship between the internal pH_i and the number of protons stored inside the vesicles at the "acidic" stage, and, therefore, determining the turnover number of H^+ATPsynthases after the pH jump [190].

As we have mentioned above, a traditional thermodynamic approach to the problem of energy coupling essentially implies that local equilibrium is set up at each stage of the energy-donating and energy-accepting processes. It means that energy released during every elementary act in the course of the energy-donating process spreads over the thermal (vibrational, rotational, and translational) degrees of freedom of the whole system. However, in the case of biopolymers we should use another approach. According to Chapters 3 and 4, an energy-transducing enzyme molecule is an enthalpic molecular machine that operates under essentially nonequilibrium conditions. For this reason, to evaluate the energy competence of certain processes ensuring the the energy-dependent step of ATP synthesis, we have to take into account the enthalpy change, ΔH, in the course of energy-donating reactions rather than the changes in free energy, ΔG. Otherwise, traditionally, the possibility of performing useful work is estimated from the comparison of free energy changes, which of necessity implies the establishment of local thermal equilibrium at every stage of the energy-transducing process.

To illustrate this, let us consider the energetics of coupling the simplest chemical reactions, i.e., the protonation/deprotonation processes, that might be relevant to energy transduction in ATPsynthases. Dissociation of a proton(s) from an enzyme acid group(s), $AH \rightarrow A^- + H^+$, may be initiated by a fast pH increase (the addition of OH^-) or, in general, by adding a "base" B^-. The net reaction of proton transfer, $AH + B^- \rightarrow A^- + BH$, can be regarded as a sum of two reactions

$$AH \rightarrow A^- + H^-, \tag{5.11a}$$

$$H^+ + B^- \rightarrow BH. \tag{5.11b}$$

According to the mass action law, the equilibrium of reaction (5.11a) would be shifted to the right in the course of reaction (5.11b). Thus, the energy-donating reaction (5.11b) supplies the energy for the energy-accepting steps of reaction (5.11a). Formation of one mole of BH from H^+ and B^- (reaction (5.11(b)) can be accompanied by the heat release corresponding to the enthalpy change, ΔH. For the reaction of the H_2O formation, $H^+ + OH^- \rightarrow H_2O$, heat production corresponds to the standard enthalpy change $\Delta H = -58.6$ kJ/mole. As we have discussed in Chapter 2, for low-molecular acids in the condensed phase the energy coupling of reactions (5.11a) and (5.11b) is indirect. The energy released in reaction (5.11b) dissipates over the macroscopic compartment (thermostat), while the energy absorbed in every act of the energy-accepting reaction (5.11a) of the AH ionization, is derived from the thermostat. For low-molecular compounds, energy initially localized on the ionized molecule A^- completely dissipates after 10^{-12}–10^{-13} s.

On the other hand, ionization of acid groups in a macromolecule (for instance, ATPsynthase) could lead to the appearance of relatively long-living nonequilibrium states. Subsequent enzyme relaxation to equilibrium can include, in general, not only the changes in the conformation, i.e., the "architecture" of the protein globule, but also the additional binding of cations that becomes possible after the ionization of an enzyme acid group. According to [54], the energy is initially conserved in the form of a mechanically strained nonequilibrium state, and then dissipates slowly in the course of a subsequent, rather slow, relaxation along a specific mechanical degree of freedom of the molecular machine. It must be emphasized again that we are now speaking not about the change in the Gibbs free energy but about a part of the total energy of a system (the so-called "stored energy," [54, 204]). It is the enthalpy change occurring in the course of the energy-donating process that ensures the performance of useful work, i.e., the liberation of the tightly-bound ATP from the coupling factor.

The amount of energy stored in a protein globule after fast deprotonation, even of a few acidic enzyme molecule groups, might appear to be quite sufficient to ensure the energy-dependent steps of ATP synthesis. For a qualitative illustration of this idea we can turn to trivial calculations of the potential energy increase immediately after the deprotonation of a protein acidic group. Let us imagine that a certain acidic group, say —COO^-, is buried in a protein globule, being at distance d from another protein-buried protonizable group, e.g., the amino group—NH_3^+. Immediately after a proton dissociation from the —NH_3^+ group (that is equivalent to the withdrawal of proton from —NH_3^+ to infinity), the potential energy of the protein globule will increase up to the value ΔW, and can be evaluated as $\Delta W = ke^2/(D_p d)$, where k is Coulon's constant, e is the electron charge, and D_p is a dielectric constant of the protein moiety. Taking $D_p \cong 5$ and $d \cong 3-5$ Å, we obtain $\Delta W \cong 55-90$ kJ/mole. Of course, this state of protein globule is energetically unfavorable. With the following protein conformational relaxation and interaction with solvent molecules, the mutual position of charged groups and their surroundings will change, thus leading to energy dissipating into heat. However, for tightly coupled systems, at least part of this energy can be employed to perform useful work. In particular, this might be the case for the local mechanism of membrane phosphorylation, when an essential part of the energy released in the course of the energy-donating processes (the enthalpic term of free energy changes) does not dissipate over the thermal degrees of freedom in the bulk phase.

Until the establishment of the real steady state, the enthalpy changes (which practically do not depend on the concentrations of reactants in the system) would determine the process direction. Experimental data also give evidence that the enthalpy changes associated with neutralization of certain functional groups in the acid-base cluster of the enzyme are high enough to be energetically competent in ensuring the energy-requiring step of ATP synthesis. Actually, the heat effects of the neutralization of basic groups may be high enough. Typical values of the ionization enthalpy for —OH, —NH_3, and imidazole groups are about $30-50$ kJ/mole [205]. Thus, using only about $25-30\%$ of the energy stored in a protein globule after deprotonation of three groups in an acid-base cluster, engaged in the process of ATP synthesis, could ensure an energetically unprofitable process of ATP liberation from the F_1 catalytic site.

Summing up, we can conclude that ATPsynthases may be considered as enthalpic molecular machines which operate in essentially nonequilibrium states formed after the ionization of several acid groups of the enzyme. *Closed membranes are necessary to provide a cyclic functioning of ATPsynthases under steady state conditions.* We hope that this approach would appear to be fruitful for the explanation of other experimental data concerning energy transduction in the living cell.

References

1. A.L. Lehninger (1964), *The Mitochondrion*, Bendjamin, New York.
2. D. Keilin (1966), *The History of Cell Respiration and Cytochrome.* Cambridge University Press, Cambridge.

3. E.C. Slater (1966), *Oxidative Phosphorylation. Comprehensive Biochemistry*, vol. 14, Elsevier, Amsterdam, pp. 327–396.

4. H.M. Calckar (1969), *Biological Phosphorylation. Development of Concepts.* Prentice-Hall, Englewood Cliffs, NJ.

5. H.A. Krebs (1970), *Perspect. Biol. Med.* **14**, 154–170.

6. F. Lipmann (1971), *Wonderings of a Biochemist*, Wiley–Interscience, New York.

7. D.I. Arnon (1977), In: *Encyclopedia of Plant Physiology*, New Series, vol. 5, Springer-Verlag, Heidelberg, pp. 7–26.

8. E. Racker (1980), *Fed. Proc.* **39**, 210–215.

9. L. Ernster and G. Schatz (1981), *J. Cell Biol.* **91**, 227–255.

10. E.C. Slater (1981), In: *Mitochondria and Microsomes* (C.P. Lee, G. Schatz, and G. Dallner, Eds.), Addison-Wesley, Reading, MA, pp. 15–43.

11. A. Tzagoloff (1982), *Mitochondria*, Plenum Press, New York.

12. J.K. Hoober (1984), *Chloroplasts*, Plenum Press, New York.

13. D.G. Nickolls (1982), *Bioenergetics. An Introduction to the Chemiosmotic Theory*, Academic Press, London.

14. V.P. Skulachev (1988), *Membrane Bioenergetics*, Springer-Verlag, Heidelberg.

15. *Photosynthesis* (1982), vol. 1 (Govindjee, Ed.) Academic Press, New York.

16. M. Wikstrom and M. Saraste (1984), In: *Bioenergetics* (L. Ernster, Ed.), Elsevier, Amsterdam, pp. 49–94.

17. Y.K. Shen and G.M. Shen (1962), *Scientica Sin.* **11**, 1097–1106.

18. G. Hind and A. Jagendorf (1963), *Proc. Nat. Acad. Sci. USA* **49**, 715–722.

19. A. Jagendorf and E. Uribe (1966), *Proc. Nat. Acad. Sci. USA* **55**, 170–177.

20. E. Racker and W. Stoeckenius (1974), *J.Biol.Chem.* **249**, 662–663.

21. W.S. Thayer and P.C. Hinkle (1975), *J. Biol. Chem.* **250**, 5330–5335.

22. D.J. Smith and P.D. Boyer (1976), *Proc. Nat. Acad. Sci. USA* **73**, 4314–4318.

23. D.J. Smith, B.O. Strokes, and P.D. Boyer (1976), *J. Biol. Chem.* **251**, 4165–4171.

24. H.T. Witt, E. Schlodder, and P. Graber (1976), *FEBS Lett.* **69**, 271–276.

25. N. Sone, M. Yoshida, H. Hiata, and Y. Kagawa (1977), *J. Biol. Chem.* **252**, 2956–2960.

26. N. Sone, Y. Takeuchi, M. Yoshida, K. Ohno, and Y. Kagawa (1977), *J. Biochem.* **82**, 1751–1758.

27. M. Rogner, K. Ohno, T. Hamamoto, N. Sone, and Y. Kagawa (1979), *Biochem. Biophyis. Res. Comm.* **91**, 362–367.

28. J. Teissie, B.E. Knox, T.Y. Tsong, and J. Wehrle (1981), *Proc. Nat. Acad. Sci. USA* **78**, 7473–7477.

29. C. Vincler and R. Korenstein (1981), *Proc. Nat. Acad. Sci. USA* **79**, 3183–3187.

30. T. Hamamoto, K. Ohno, and Y. Kagawa (1982), *J. Biochem.* **91**, 1759–1766.

31. R.P. Hangarter and N.E. Good (1982), *Biochim. Biophys. Acta* **682**, 381–404.

32. P. Graber, U. Junesch and G.H. Schatz (1984), *Ber. Bugsenges. Phys. Chem.* **88**, 599–608.

33. G. Schmidt and P. Graber (1985), *Biochim. Biophys. Acta* **808**, 46–51.

34. U. Junesch and P. Graber (1985), *Biochim. Biophys. Acta* **809**, 429–434.

35. G. Schmidt and P. Graber (1985), *Biochim. Biophys. Acta* **890**, 392–394.

36. E.C. Slater (1953), *Nature* **172**, 975–978.

37. B. Chance and G.R. Williams (1956), *Adv. Enzymol.* **17**, 65–134.

38. P.D. Boyer (1965), In: *Oxidases and Related Systems* (T.E. King, H.S. Mason and M. Morrison, Eds.) vol. 2, Wiley, New York. pp. 994–1008.

39. P. Mitchell (1961), *Nature* **191**, 144–148.

40. P. Mitchell (1966), *Chemiosmotic Coupling in Oxidative and Photosynthetic Phosphorylation*, Glynn Research, Bodmin, UK.
41. R.J.P. Williams (1961), *J. Theoret. Biol.* **1**, 1–13.
42. R.J.P. Williams (1962), *J. Theoret. Biol.* **3**, 209–229.
43. R.J.P. Williams (1975), *FEBS Lett.* **53**, 123–125.
44. R.J.P. Williams (1978), *Biochim. Biophys. Acta* **505**, 1–44.
45. P. Mitchell (1977), *Ann. Rev. Biochem.* **46**, 996–1005.
46. P. Mitchell (1974), *FEBS Lett.* **43**, 189–194.
47. Y. Kagawa (1984), In: *Bioenergetics* (L. Ernster, Ed.), Elsevier, Amsterdam, pp. 149–186.
48. H.V. Westerhoff, B.A. Melandry, G. Venturoli, G.F. Azzone, and D.B. Kell (1984), *Biochim. Biophys. Acta* **768**, 257–292.
49. S.J Ferguson (1985), *Biochim. Biophys. Acta* **811**, 47–95.
50. H.V. Westerhoff and K. van Dam (1987), *Thermodynamics and Control of Biological Energy Transduction*, Elsevier/North Holland, Amsterdam.
51. J.W. Stucki (1980), *European J. Biochem.* **109**, 269–283.
52. L.A. Blumenfeld and V.A. Koltover (1972), *Mol. Biol. (USSR)* **6**, 161–166.
53. L.A. Blumenfeld (1981), *Problems of Biological Physics*, Springer-Verlag, Heidelberg.
54. L.A. Blumenfeld (1983), *Physics of Bioenergetic Processes*, Springer-Verlag, Heidelberg.
55. W. Auslander and W. Junge (1975), *FEBS Lett* **59**, 310–315.
56. W. Junge, W. Auslander, A.J. McGeer, and T. Runge (1979), *Biochim. Biophys. Acta* **546**, 121–141.
57. W. Junge (1982), *Curr. Top. Membr. Transp.* **16**, 431–463.
58. V. Forster, Y.Q. Hong, and W. Junge (1981), *Biochim. Biophys. Acta* **638**, 141–152.
59. Y.Q. Hong and W. Junge (1983), *Biochim. Biophys. Acta* **722**, 197–208.
60. J.K.M. Roberts and O. Jardetzky (1981), *Biochim. Biophys. Acta* **639**, 53–76.
61. *Intracellular pH: Its Measurement, Regulation and Utilization in Intracellular Function* (R. Nuccitelli and D.W. Deamerr, Eds.), Liss, New York, 1982.
62. G.F. Azzone, D. Pietboron, and M. Zoratti (1984), In: *Current Topics in Bioenergetics*, vol. 13, pp. 1–77.
63. H. Rottenberg, T. Grunwald, and M. Avron (1971), *FEBS Lett.* **13**, 41–44.
64. S. Schuldiner, H. Rottenberg, and M. Avron (1972), *European J. Biochem.* **25**, 64–70.
65. H. Rottenberg (1979), *Meth. Enzymol.* **55**, 547–569.
66. A.T. Quintanihla and R.J. Mehlholrn (1978), *FEBS Lett.* **91**, 161–165.
67. A.N. Tikhonov and L.A. Blumenfeld (1985), *Biophysics (USSR)* **30**, 527–537.
68. U. Pick and M. Avron (1976), *FEBS Lett.* **65**, 348–353.
69. D. Siefermann-Harms (1978), *Biochim. Biophys. Acta* **504**, 265–277.
70. M. Masarova and A.N. Tikhonov (1988), In: *Structural Dynamics of Photobiological Membranes and Receptory Processes* (I.D. Volotovsky, Ed.), Minsk, USSR, p. 87.
71. B. Rumberg and U. Siggel (1969), *Naturwissenschaften* **56**, 130–132.
72. W. Haehnel (1976), *Biochim. Biophys. Acta* **440**, 506–521.
73. W. Haehnel (1984), *Ann. Rev. Plant Physiol.* **35**, 659–693.
74. A.N. Tikhinov, G.B. Khomutov, and E.K. Ruuge (1984), *Photobiochem. Photobiophys.* **8**, 261–269.

75. R. Mitchell, A. Spillmann, and W. Haehnel (1990), *Biophysical J.* **58**, 1011–1024.
76. A.N. Tikhonov, G.B. Khomutov, E.K. Ruuge, and L.A. Blumenfeld (1981), *Biochim. Biophys. Acta* **637**, 321–333.
77. A.N. Tikhonov and A.A. Timoshin (1985), *Biol. Membranes (USSR)* **2**, 349–362.
78. A.N. Tikhonov and A.A. Timoshin (1985), *Biol. Membranes (USSR)* **2**, 608–626.
79. S.P. Berg and D.M. Nesbitt (1980), *FEBS Lett.* **112**, 101–104.
80. D.M. Nesbitt and A.S. Berg (1980), *Biochim. Biophys. Acta* **593**, 353–361.
81. D.M. Nesbitt and A.S. Berg (1982), *Biochim. Biophys. Acta* **679**, 169–174.
82. L.A. Blumenfeld, R.M. Davydov, and A.N. Tikhonov (1989), *J. Molc. Liq.* **42**, 231–253.
83. U. Pick, H. Rottenberg, and M. Avron (1973), *FEBS Lett.* **32**, 91–94.
84. U. Pick, H. Rottenberg, and M. Avron (1973), *FEBS Lett.* **48**, 32–36.
85. A.R. Portis and R.E. McCarty (1974), *J. Biol. Chem.* **249**, 6250–6254.
86. W.S. Chow and A.B. Hope (1976), *Austral. J. Plant. Physiol.* **3**, 141–142.
87. A.N. Tikhonov and A.V. Shevyakova (1985), *Biol. Membranes (USSR)* **2**, 776–788.
88. S.G. Gilmiarova, M. Masarova, and A.N. Tikhonov (1985), *Biophysics (USSR)* **30**, 709–710.
89. L.A. Staehelin, P.A. Armond, and K.R. Miller (1976), *Brookhaven Symp. Biol.* **28**, 278–315.
90. P.A. Armond, L.A. Staehelin, and C.J. Arntzen (1977), *J. Cell Biol.* **73**, 400–418.
91. K.R. Miller and R.A. Gusman (1979), *Biochim. Biophys. Acta* **546**, 481–497.
92. K.R. Miller (1980), *Biochim. Biophys. Acta* **592**, 143–152.
93. B. Andersson and J.M. Anderson (1980), *Biochim. Biophys. Acta* **593**, 427–440.
94. J.M. Anderson (1982), *FEBS Lett.* **138**, 62–66.
95. J.M. Anderson and W. Haehnel (1982), *FEBS Lett.* **146**, 13–17.
96. J.M. Anderson and R. Malkin (1982), *FEBS Lett.* **148**, 293–296.
97. J.M. Anderson and A. Melis (1983), *Proc. Nat. Acad. Sci. USA* **80**, 745–749.
98. W.S. Chow, C. Miller, and J.M. Anderson (1991), *Biochim. Biophys. Acta* **1057**, 69–77.
99. C. Sigalat, F. Haraux, F. de Kouchkovsky, S.P.N. Hung, and Y. de Kouchkovsky (1985), *Biochim. Biophys. Acta* **809**, 403–413.
100. F. Haraux, C. Sigalat, A. Morau, and Y. de Kouchkovsky (1983), *FEBS Lett.* **155**, 248–251.
101. R.P. Hangarter, R.W. Jones, D. Ort, and J. Whitmarsh (1987), *Biochim. Biophys. Acta* **890**, 106–115.
102. H. H. Stiehl and H.T. Witt (1969), *Z. Naturforsch.* **B24**, 1588–1598.
103. H. T. Witt (1971), *Quart. Rev. Biophys.* **4**, 365–477.
104. H. T. Witt (1979), *Biochim. Biophys. Acta* **505**, 355–427.
105. A. Polle and W. Junge (1986), *Biochim. Biophys. Acta* **848**, 257–264.
106. W. Junge and A. Polle (1986), *Biochim. Biophys. Acta* **848**, 265–273.
107. W. Junge and S. McLaughlin (1987), *Biochim. Biophys. Acta* **890**, 1–5.
108. D. Ort (1978), *European J. Biochem.* **85**, 479–485.
109. Y. de Kouchkovsky and F. Haraux (1981), *Biochem. Biophys. Res. Comm.* **99**, 205–212.
110. M. Renganathan, R.S. Pan, R.G. Ewy, S.M. Theg, F.C.T. Allnutt, and R.A. Dilley (1991), *Biochim. Biophys. Acta* **1059**, 16–27.
111. J. Barber (1980), *Biochim. Biophys. Acta* **594**, 253–308.

112. R. Tiemann and H.T. Witt (1982), *Biochim. Biophys. Acta* **681**, 235–247.

113. R.A. Dilley, S.M. Theg, and W.A. Beard (1987), *Ann. Rev. Plant Physiol.* **38**, 347–389.

114. D.B. Kell (1979), *Biochim. Biophys. Acta* **549**, 55–99.

115. D.F. Wilson and N.S. Forman (1982), *Biochemistry* **20**, 1438–1444.

116. D.Nickolls (1984), In: *Bioenergetics* (L. Ernster, Ed.), Elsevier, Amsterdam, pp. 29–49.

117. D.B. Kell and H.T. Westerhoff (1985), In: *Organized Multienzyme Systems; Catalytic Properties* (G.R. Welch, Ed.), Academic Press, New York, pp. 63–193.

118. D.R. Ort and R.A. Dilley (1976), *Biochim. Biophys. Acta* **449**, 95–107.

119. D.R. Ort and R.A. Dilley (1976), *Biochim. Biophys. Acta* **449**, 108–124.

120. T. Graan, S. Flores, and D.R. Ort (1981), In: *Energy Coupling in Photosynthesis* (B.R. Selman and S. Reimer-Selman, Eds.), Elsevier/North-Holland, Amsterdam, pp. 25–34.

121. W.A. Beard and R.A. Dilley (1986), *FEBS Lett.* **210**, 57–62.

122. G. Chiang and R.A. Dilley (1987), *Biochemistry* **26**, 4911–4916.

123. C. Sigalat, F. Haraux, F. de Kouchkovsky, S.P.N. Hung, and Y.de Kouchkovsky (1985), *Biochim. Biophys. Acta* **809**, 403–413.

124. V.K. Opanasenko, T.P. Redko, V.P. Kuzmina, and L.S. Yaguzhinsky (1985), *FEBS Lett.* **187**, 257–269.

125. J.A. Laszlo, G.M. Baker, and R.A. Dilley (1984), *Biochim. Biophys. Acta* **764**, 160–169.

126. S.M. Theg, G. Chiang, and R.A. Dilley (1988), *J. Biol. Chem.* **263**, 673–681.

127. S.M. Theg and P.H. Homann (1982), *Biochim. Biophys. Acta* **679**, 221–234.

128. S.M. Theg, J.D. Johnson, and P.H. Homann (1982), *FEBS Lett.* **145**, 25–29.

129. S.M. Theg and W. Junge (1983), *Biochim. Biophys. Acta* **679**, 294–307.

130. J.A. Laszlo, G.M. Baker, and R.A. Dilley (1984), *J. Bioenerg. Biomembr.* **16**, 37–51.

131. Y. Kagawa (1984), *J. Biochem.* **95**, 295–298.

132. H. Strotmann and S. Bickel-Sandkotter (1984), *Ann. Rev. Plant Physiol.* **35**, 97–120.

133. Y. Hatefi (1985), *Ann. Rev. Biochem.* **54**, 1015–1069.

134. P.V. Vignais and J. Lunardi (1985), *Ann. Rev. Biochem.* **54**, 977–1014.

135. E. Schneider and K. Altendorf (1987), *Microbiol. Rev.* **51**, 477–497.

136. C.M. Nalin and N. Nelson (1987), *Curr. Topics Bioenerg.* **15**, 273–294.

137. P.D. Boyer (1987), *Biochemistry* **26**, 8503–8507.

138. P.L. Pedersen and E. Carafoli (1987), *Trends Biochem. Sci.* **12**, 146–150; 186–189.

139. A.E. Senior (1988), *Physiol. Rev.* **68**, 177–231.139.

140. R.L. Cross (1988), *J. Bioenerg. Biomembr.* **20**, 395–405.

141. J.H. Wang (1988), *J. Bioenerg. Biomembr.* **20**, 407–422.

142. X. Ysern, L.M. Amzel, and P.L. Pedersen (1989), *J. Bioenerg. Biomembr.* **20**, 423–450.

143. D.C. Gautheron and C. Godinot (1988), *J. Bioenerg. Biomembr.* **20**, 451–468.

144. M. Futai, T. Noumi and M. Maeda (1988), *J. Bioenerg. Biomembr.* **20**, 469–480.

145. A. Matsuno-Yagi and Y. Hatefi (1988), *J. Bioenerg. Biomembr.* **20**, 481–502.

146. H. Tiedge and G. Schafer (1989), *Biochim. Biophys. Acta* **977**, 1–9.

147. G. Forti, L. Rosa, and F. Garlaschi (1972), *FEBS Lett.* **27**, 23–26.

148. N. Yamamoto, S. Yoshimura, T. Higuti, K. Nishikawa, and T. Norio (1972), *J. Biochem. Tokyo* **72**, 1397–1406.

149. T. Oku, K. Hosoi, G. Soe, T. Kakuno, and T. Norio (1974), *J. Biochem. Tokyo* **76**, 223–236.

150. D.A. Harris, J. Rosing, R.J. Van de Stadt, and E.C. Slater (1973), *Biochim. Biophys. Acta* **314**, 149–153.

151. D.A. Harris and E.C. Slater (1975), *Biochim. Biophys. Acta* **387**, 335–348.

152. J.H. Joung, E.F. Korman, and J. McClick (1971), *Bioenorg. Chem.* **3**, 1–15.

153. P. Mitchell (1985), *J. Bioenerg.* **3**, 5–24.

154. J.R. Knowles (1980), *Ann. Rev. Biochem.* **49**, 877–919.

155. M. Kohn (1982), *Ann. Rev. Biophys. Bioenerg.* **11**, 23–42.

156. M.R. Webb, C. Grubmeyer, H.S. Penefsky, and D.R. Trentham (1980), *J. Biol. Chem.* **255**, 11637–11639.

157. P. Senter, F. Eckstein, and Y. Kagawa (1983), *Biochemistry* **22**, 5514–5518.

158. N. Williams and P.C. Collemans (1982), *J. Biol. Chem.* **257**, 2834–2841.

159. J. Rosing, C. Kayalar, D.J. Smith, and P.D. Boyer (1976), *Biochem. Biophys. Res. Comm.* **72**, 1–8.

160. G. Rosen, M. Gresser, C. Vincler, and P.D. Boyer (1977), *J. Biol. Chem.* **254**, 10654–10661.

161. P.D. Boyer, B.O. Strokes, R.G. Wolcott, and C. Degani (1975), *Federation Proc.* **34**, 1711–1717.

162. C. Kayalar, J. Rosing, and P.D. Boyer (1976), *Biochem. Biophys. Res. Comm.* **72**, 1153–1159.

163. E.C. Slater (1974), In: *Dynamics of Energy Transducing Membranes* (L. Ernster, R.W. Estabrook and E.C. Slater, Eds.), Elsevier, Amsterdam, pp. 1–20.

164. J. Rosing and E.C. Slater (1972), *Biochim. Biophys. Acta* **267**, 275–290.

165. G.L. Chaote, R.L. Hutton, and P.D. Boyer (1979), *J. Biol. Chem.* **254**, 286–290.

166. R.L. Hutton and P.D. Boyer (1979), *J. Biol. Chem.* **254**, 9990–9993.

167. P.D. Boyer and W.E. Kohlbrenner (1981), In: *Energy Coupling in Photosynthesis* (B.R. Selman and S. Reimer-Selman, Eds.), Elsevier/North-Holland, Amsterdam, pp. 231–240.

168. G. Weber (1975), *Adv. Protein Chem.* **29**, 1–83.

169. H.M. Younis, G. Weber, and J.S. Boyer (1983), *Biochemistry* **22**, 2505–2512.

170. I.J. Ryrie and A.T. Jagendorf (1972), *J. Biol. Chem.* **247**, 4453–4457.

171. R. Kraayenhof and E.C. Slater (1974), *Proc. 3rd Int. Congr. Photosynth* (M. Avron, Ed.), vol. 2, Elsevier, Amsterdam, pp. 985–996.

172. R. Kraayenhof (1977), In: *Current Topics in Bioenergetis* (D. Rao Sanadi, Ed.), Academic Press, New York, pp. 423–428.

173. R. Wagner and W. Junge (1980), *FEBS Lett.* **114**, 327–333.

174. R. Wagner, N. Carrillo, and W. Junge (1981), *FEBS Lett.* **136**, 208–212.

175. R. Wagner and W. Junge (1982), *Biochemistry* **21**, 1890–1899.

176. S. Ohta, M. Tsuboi, T. Ohshima, M. Yoshida, and Y. Kagawa (1980), *J. Biochem.* **87**, 1609–1617.

177. T. Gantchev, T. Tsanova, G. Gotchev, and V. Timofeev (1990), *Biochim. Biophys. Acta* **1015**, 69–78.

178. R.I. Feldman and D.S. Sigman (1982), *J. Biol. Chem.* **257**, 1676–1683.

179. J. Sakamoto and Y. Tonomura (1983), *J. Biochem.* **93**, 1601–1614.

180. M. Yoshida (1983), *Biochem. Biophys. Res. Comm.* **114**, 907–912.

181. E.N. Moudrianakis and M. Tieffert (1976), *J. Biol. Chem.* **251**, 7796–7801.

182. I.V. Malenkova, S.P. Kuprin, R.M. Davydov, and L.A. Blumenfeld, (1982), *Biochim. Biophys. Acta* **682**, 179–183.

183. V.A. Serezhenkov, I.V. Malenkova, Sh.T. Talybov, A.S. Kaprelyants, R.M. Davydov, and L.A. Blumenfeld (1986), *Biophysics (USSR)* **31**, 972–975.
184. L.A. Blumenfeld (1987). In: *Structure, Dynamics and Function of Biomolecules* (A. Ehrenberg, R. Rigler, A. Graslund, and L. Nilson, Eds.), Springer-Verlag, Berlin, pp. 187–190.
185. L.A. Blumenfeld, I.V. Malenkova, S.S. Kormer, V.A. Serezhenkov, E.I. Mileykovskaya, and R.M. Davydov (1986), *Proc. Acad. Sci. USSR* **288**, 1494–1496.
186. L.A. Blumenfeld, M.G. Goldfield, V.D. Mikoyan, and I.S. Solovyev (1987), *Molec. Biol. (USSR) Molec. Biol.* **21**, 323–329.
187. S.B. Ryzhikov, L.A. Blumenfeld, Yu.M. Petrusevich, and A.N. Tikhonov (1990), *Biophysics (USSR)* **35**, 141–142.
188. L. De Meis and R.H. Tume (1977), *Biochemistry* **16**, 4455–4463.
189. L. DeMeis and G. Inesi (1982), *J. Biol. Chem.* **257**, 1289–1294.
190. L.A. Blumenfeld and A.N. Tikhonov (1987), *Biophysics (USSR)* **32**, 800–813.
191. L.A. Blumenfeld (1988), *J. Biomol. Structure Dynamics* **6**, 23–33.
192. L.A. Blumenfeld, S. Greschner, M.V. Genkin, R.M. Davydov, and N.M. Roldugina (1976), *Studia Biophysica.* **57**, 110.
193. R. Vallejos (1981), In: *Energy Coupling in Photosynthesis* (B.R. Selman and S. Reimer-Selman, Eds.), Elsevier/North-Holland, Amsterdam, pp. 129–139.
194. C. Carmeli, Y. Hochman, and Y. Lifshitz (1981), In: *Energy Coupling in Photosynthesis* (B.R. Selman and S. Reimer-Selman, Eds.), Elsevier/North-Holland, Amsterdam, pp. 111–117.
195. I.A. Kozlov (1981), In: *Chemiosmotic Proton Circuits in Biological Membranes* (V.P. Skulachev and P.C. Hinkle, Eds.), Addison-Wesley, Reading, MA, pp. 407–420.
196. R.P. Magnuson and R.E. McCarty (1976), *J. Biol. Chem.* **251**, 6874–6877.
197. R.E. McCarty, J.S. Fuhrmann, and M. Tsuchia (1971), *Proc. Nat. Acad. Sci. USA* **68**, 2522–2526.
198. J.M. Gould (1976), *FEBS Lett.* **66**, 312–316.
199. P. Graber, M. Burmeister, and M. Hortsch (1981), *FEBS Lett.* **136**, 25–31.
200. S.G. Gilmiarova, A.N. Tikhonov, E.K. Ruuge, and L.A. Blumenfeld (1985), *Proc. Acad. Sci. USSR* **283**, 1400–1403.
201. R. Wagner, G. Ponse, and H. Strotmann (1986), *European J. Biochem.* **161**, 205–209.
202. A.D. Vinogradov (1984), *Biochemistry (USSR)* **49**, 1220–1238.
203. A. Brune, J. Spillecke, and A. Kroger (1987), *Biochim. Biophys. Acta* **893**, 499–507.
204. C.W.F. McClare (1971), *J. Theoret. Biol.* **30**, 1–34.
205. J.T. Edsall and H. Guttfruend, *Biothermodynamics. The Study of Biochemical Processes at Equilibrium*, Wiley, Chichester.
206. V.A. Tverdislov, A.N. Tikhonov, and L.V. Yakovenko (1987), *Physical and Chemical Mechanisms of Biological Membranes Functioning* (in Russian), Moscow University Press, Moscow.
207. A.K. Kukushkin and A.N. Tikhonov (1988), *Lectures on Photosynthesis of Higher Plants* (in Russian), Moscow University Press, Moscow.
208. S. Engelbrecht and W. Junge (1990), *Biochem. Biophys. Acta* **1015**, 379–390.

Afterword

The crucial word in the title of this book is *machines*. The adjective *molecular* states that we were considering here a special class of machines determining some important aspects of the functioning of the living cell. Our discussion has been restricted to molecular machines responsible for certain enzyme reactions and bioenergetic processes. All conclusions reached in the preceding chapters are applicable to molecular machines participating in other biological processes, such as protein biosynthesis, the storage and transfer of hereditary information, visual perception, etc.

It is not yet common knowledge that the existence of machines is profoundly connected with life, and with biology. *There are no machines in nonliving nature except those made by living objects*. All machines are either parts of living creatures, or are made by living creatures.

What is a machine? A machine is a meaningful construction possessing a goal, and able to perform the directional transfer of energy and/or information. To be a machine is rather more sophisticated than to be just a construction. According to the above definition, the obligatory attribute of a machine is its meaning. When we are speaking of a machine we can ask the question "*Why?*".

Meaning is a teleological concept. To have a meaning is to have a purpose. For a man-made machine, the purpose is determined by its designer. For living objects, the purpose has been "settled" in the course of biological evolution: the ultimate goal is the survival of the species. Considering, say, the properties of a sodium chloride crystal, we cannot ask: "Why does this crystal have cubic symmetry?". We can only put the question "*Why?*", and get the answer: "Because such a structure corresponds to the minimal value of energy." However, we can ask: "Why are the amino acid residues in a γ-globulin molecule positioned in this specific orderly way?", and get the answer: "For the purpose of ensuring the specific immunological activity of this protein." For biological structures the meaning assumes a particular significance, since it determines the importance of certain chemical and physical properties of these structures. This problem was discussed in more detail by one of us

(Blumenfeld, L.A. *Problems of Biological Physics*, Spinger-Verlag, Heidelberg, 1981).

It has become fashionable today to speak of the machine-like behavior of enzymes, intracellular particles (e.g., ribosomes), etc., during their functioning. The phrases "a protein is a machine," "an enzyme is a machine" are now trivial cliches, and at the same time remain vague. The main reason for this is the very approach used by the majority of scientists in the treatment of the chemical properties of biopolymers. In spite of speculation regarding the "machineness" of proteins, they apply, as a rule, to the conventional approaches of chemical thermodynamics and chemical kinetics that have been developed for the reactions of low-molecular compounds in gaseous phases and dilute solutions. These approaches are based essentially on the classical statistical physics of ergodic systems, i.e., on the assumption that the systems under consideration have only statistical, thermal degrees of freedom fast enough to exchange energy for each other. However, if biological constructions (beginning at the level of macromolecules) are machines, in the course of their functioning there might be excited specific, mechanical degrees of freedom which exchange slowly with the thermal ones. This requires an essentially different approach to their description.

The most important intracellular constructions are not simply machines, they are *very small machines*. This circumstance determines new peculiarities in their behavior, as well as introducing new features into the scientific analysis of the functioning of molecular machines. In this book we have tried to discuss some principal aspects of this problem by analyzing the processes of intracellular energy transduction.

Index

ATP   17
   synthesis of   18, 112, 116
      by acid-base transition   146
      conditions for steady state   159
      elementary act of   144, 151
      energy-requiring step in   146
      local mechanism of   112
      model of   159
      relaxation concept of   154
ATPsynthase   113
   cyclic functioning of   159
   structure of   144

Biomembranes   45
   energy transduction in   112

Chemical affinity   11
Chemical reactions   6
   channeled   73
   in small vesicles   60, 64
   kinetics of   7
      temperature dependence of   13, 19
   mass action law for   6, 68
   thermodynamics of   10
Chemiosmotic concept   112
Chloroplasts   61
   structure of   62
Coupling factor   61, 151

Electrochemical proton gradient   117
Electron transport chain   112
Energy   4
   free   10
   stored   39
Energy coupling   20, 112
   in biomembranes   116

in chemical reactions   20
   conversion factor   26
   efficiency of   26
   indirect mechanism of   21
      enthalpic mechanism of   22
      entropic mechanism of   24, 30
Enzyme catalysis   86
   earlier theories of   89
   relaxation concept of   94

Fluctuations   64

Machines   2
   macroscopic   45
   molecular   2, 48
      models of   52
Membrane phosphorylation   116
Mitochondria   18
   structure of   113, 115

Oxidative phosphorylation   162

Photophosphorylation   162
Proton transfer   103, 108, 133, 140

Relaxation concept   95

Spin labels   122

Thylakoid   113
Transmembrane proton gradient   121
   measurements of   123, 127
   "threshold" value of   163